AFGHANISTAN

THE END OF THE U.S. FOOTPRINT AND THE RISE OF THE TALIBAN RULE

Afghanistan: The End of the U.S. Footprint and the Rise of The Taliban Rule is published under Erudition, a sectionalized division under Di Angelo Publications, Inc.

Erudition

Erudition is an imprint of Di Angelo Publications.

Printed in the United States of America.

Di Angelo Publications
4265 San Felipe #1100
Houston, Texas 77027

Library of Congress
Afghanistan: The End of the U.S. Footprint and the Rise of The Taliban Rule
ISBN: 978-1-955690-24-9

Words: Hollie McKay
Photographs: Jake Simkin
Cover Design: Savina Deianova
Interior Design: Kimberly James
Editors:

For educational, business, and bulk orders, contact sales@diangelopublications.com.

1. History --- Wars & Conflicts --- Afghan War (2001-2021)
2. History --- Modern --- 21st Century
3. Political Science --- Human Rights
4. Political Science --- International Relations
5. Political Science --- Terrorism

AFGHANISTAN

THE END OF THE U.S. FOOTPRINT AND THE RISE OF THE TALIBAN RULE

WORDS BY

HOLLIE MCKAY

PHOTOS BY

JAKE SIMKIN

PHOTOS & STORIES

INTRODUCTION

PART ONE: BEFORE THE FALL

PART TWO: EARLY DAYS OF THE EMIRATE

PART THREE: NEW, OLD WAYS

PART FOUR: AFGHANISTAN FOREVER

CONCLUSION

INTRODUCTION

Our world brims with entrancing, complicated places. But none that quite hold the heart and clutch the imagination like Afghanistan, which means "Land of the Afghans."

It is a landlocked trove of timeworn beryl stones and hidden treasures, of dusty paths encrusted from the historic Silk Road, of black skies with stars that light up the serrated mountain ranges, and mud-walled homes where those with pinched faces and warm eyes eat by candles and moonlight.

Every child born in Afghanistan is a child who comes into life a victim of war. And every parent too often puts those they love most to sleep with the uncertainty that one or all may not rise to the shards of peeping pink light again.

It is the place where we journalists venture and immediately fall in love with open Afghan hands, cracked by tireless work and unwavering hospitality. It is also the berth we go to have our souls shredded into a million pieces every time smoke plumes rise into the muggy sky from yet another bomb, or when a woman weeps from behind a sea-blue burka, unable to feed the baby in her arms.

Afghanistan embraces us and destroys us. Still, we cannot stay away from its gentle call. It is a nation that has long been burned and entombed by outside occupation, by conflict, by poverty, by its own past.

Afghanistan's flaxen plains and clear blue skies have seen too much death. Too many men have taken their last breath while crouched in trenches; too many mothers have gone to the market for honeyed sweets and never come home. Too many Afghans from too many sides have lost lives and limbs in battles of ideology, power, and land.

You can forget it is a beautiful country. A beautiful country beneath the bloodshed, stitched and ripped, stitched and ripped, throughout decades of ceaseless conflict. Let us not forget that Afghanistan is also a country personified by survival, toughness, tender faces, and a zest for something better.

Curling around the famous Hindu Kush, Afghanistan remains one of the poorest places on the planet, yet so rich in its generosity. It is the snuggled land where strangers offer you their last piece of bread; lyrical souls guide you through the narrow, clay-caked streets; and time spent fantasizing of something better is always time well spent.

With the Taliban having suddenly resumed power on the sultry Sunday afternoon of August 15, 2021, there was an immediate kinesthesia of the old Afghan world passing, and the new beginning with much trepidation and fear.

Nobody in Afghanistan really wants war. Yet war always waits in the grimy shadows of broken things.

Nobody can predict what will happen next.

But life, including the book of Afghan life, overflowing with chapters of loss and love and faith and hope, goes on.

Throughout this collection, my photographer Jake and I take you on a journey through the country that has shaped our journalistic lives; has taught us what it is to suffer and what it is to overcome. We guide you through life in Afghanistan during the U.S. occupation, the dizzying government fall, the Taliban ascent amid the chaotic U.S. withdrawal, and into this new chapter being forcefully written by men from the mountains who wear their thick black turbans like crowns.

We accompany you across multiple provinces of Afghanistan in the immediate months after the Taliban, officially termed "The Islamic Emirate of Afghanistan," took hold. There are the tender moments: the afternoons filled with children begging for soccer balls and screeching with delight through the antiquated village roads; the emerald lakes iced into absolute stillness; the devout absorbed in prayers on mats by the roadside, adhering the first call to prayer against hints of light behind the ridges.

Then there is the callous side, the war-wracked sentiment, the notion that a Band-Aid was ripped from the bullet wound. The fear for what comes next, the longing many Afghans have to leave, the unraveling humanitarian catastrophe, the hunger pains fused with the ache of abandonment.

With the Taliban at the helm, there is much we know, and much that we do not.

However, we must never forget a little girl's dream to be more than an obscured outline banished to a basement and permitted only to see her world from the netting of a burqa, her existence buried beneath layers of undulating material.

During the Taliban's rule of the nation between 1996–2001—until they were unceremoniously usurped from power by the United States in the aftermath of September 11—girls' education was stopped at the secondary level. Their public lives were suddenly severed short. It was clear most would never emerge with a face that could be kissed by sunlight until they were in the twilight of their lives.

The Taliban of 2021 pledges they have changed. In many ways, they have—yet almost immediately into their tenure, schooling for girls stopped, and many women are no longer permitted to go to work if it entails collaboration with a male co-worker.

In those first confusing months, education was not the only thing high school—and some university—girls lost. Tangled within that declaration, they lost their right to a future, to some semblance of autonomy, to ambitions beyond their humble homes.

I have walked through the Ministry of Education—a run-down building with faded white walls and small tracts of greenery rising from cracks in the concrete—multiple times since the Taliban took control of Afghanistan in mid-August, 2021. The floors and stuffy office rooms are filled with the new government's most devout, as evidenced by their long beards and constant twisting of prayer beads, but I have yet to spot a single woman inside the busy halls.

It is a sober reminder of the state of girls' education inside an Afghanistan in limbo, with next to no input from the gender suffering most amid the Taliban takeover. These are the things so many of us take for granted and the things we cannot forget when it comes to fighting for Afghanistan.

On one occasion in Khost, I sit on the floor sipping tea with brave-faced older women and widows who speak passionately about the consequential necessity of female teachers and students. One of the ladies' small granddaughters—just nine years old—slips quietly beside me onto the cool concrete. She says nothing, her innocent, frightened face and milky eyes staring inquisitively into mine.

I learn that she is already engaged. Her poor parents, made even poorer by the economic crisis gripping the beleaguered country, felt they had no other means but to essentially give her away to an older man in exchange for fast funds, as per the cultural custom that the brides' families are paid.

The wedding will take place after her first period. There is no more education for the timid girl with milky brown eyes; she no longer has a choice. This is not new in Afghanistan, but with the drastic cash shortage, I fear it will become even more commonplace.

Somehow, we must let go of the past and remember it at the same time. We must embrace the reality that is, not the reality we want to be, and do what we can to be a helping hand.

Tens of thousands of young Afghan soldiers, in the prime of their warring lives, had tried to fight for a country that did not fight for them. Their president, Ashraf Ghani, escaped Afghanistan without a word, freeing the Taliban to take the presidential palace without firing a shot.

But through the course of the two-decade war, many shots—too many shots—ended lives and rendered an entire generation deprived of a father figure.

We look forward. We pray. We hope. We must.

In no country have I ever seen so many blood stains and such desperation. I have also never been to a place splitting at the seams with poetic voices and dancing eyes.

Just before the U.S. made its swan song in the summer of 2021, we returned to Afghanistan to tell the human stories and show the true costs of war.

TIE DOWN
TIE DOWN
2HH002

We can pontificate all day about statistics and percentages and death tolls and displacement numbers, easily forgetting that numbers have names and names have faces.

Each face has wrinkles that make up a map filled with tragedies to tell.

These are the ordinary Afghans who have lost someone they loved most in this life. These are the ordinary Afghans who have no choice but to endure decades of hardship and havoc. These are the ordinary Afghans who supported the United States through two decades of occupation, who nursed the wounded and sacrificed time with their own families for weeks, or months, in the pursuit of freedom and the backing of the American invasion.

Unlike many other conflict-drenched countries I have canvassed, where people work ceaselessly to survive and struggle and yet don't know that life could be any different or better, Afghans have tasted something else. Something different, something more.

Everyone in Afghanistan has lost someone.

Everyone is a survivor. Everyone has a strange story to tell, usually passed on offhandedly as we eat bread and oily rice with our bare hands from plastic mats, our backs against exquisitely patterned cushions.

There is the Mujahideen commander who told me casually about his encounter with Usama bin Ladin at an orange farm in Jalalabad in 1989. There is the witch doctor who tried to heal my woeful food poisoning by flapping his hands.

And I will not forget the women who boiled water and lathered my hair with soap when there was no electricity or running water. I will not forget the tribal elder who excitedly demonstrated the best way to fold my new checkered, silk hijab as the Taliban waited to take me away.

When I go to sleep most nights, I wonder what is, and what will become, of the flimsy Afghan dream born in the time of the American-led war.

Two days before much of the nation's north fell to Taliban rule, I met a thin-boned, bubbly young Afghan man by the name of Noorzai Ibrahimi. He is one of Afghanistan's top skateboarders—having gone from a beggar boy in the streets offering to weigh people with his cracked scale to learning how to wheel and flip and fly.

Noorzai is a long and dear friend of Jake's, who had plucked him from child poverty and taught him to skateboard years ago. He is twenty and madly in love. He laughs at the end of every sentence.

But in the height of the maddened airport evacuation, Noorzai had to flee his homeland, his family, and his desires, or else his beloved boarding would have been rendered a thing of the past.

Small things. But small things matter inside the Afghan dream.

Everyone wants the bombing and battery to stop. The Taliban says the war is over, yet we cannot see the peace. Is she there?

Is war ever really over? Or is this just the beginning of another war, another epoch of agony?

PART ONE:

BEFORE THE FALL

I look out at the dried-out fragments of a nation burned alive by loathing, at Afghans going through the motions of living like ghosts teetering on the threshold of a ticking time bomb.

Waiting for the explosion. The one that could come at any time.

Many in Kabul believe that the Taliban cannot grasp the capital, the ultimate prize sandwiched in the plains, safeguarded by the undulating mountains.

The frail foreign protection that they had come to know over two decades is now the rope burning in their palms, slipping away faster and faster. Soon, they will have no alternative but to let go.

Everyone knows it. But nobody wants to say. So therein still lies a sliver of hope that Americans will change their mind and stay—stay as the feeble buffer even if it protracts the war. The other side, their absence, the unknown, feels far more uncomfortable.

The giant Afghan flag flutters in the late summer winds atop Wazir Akbar Khan Hill, stark rectangles of green, red, and black. It sways above the many embassies, the presidential palace, the NATO Resolute Support Mission, and the foreign fortresses.

The flaps of blurred color, lonely on the mountaintop, are dying in the last streaks of daylight.

Yet here the people of Kabul are still living. There is no other option. Warm bread wafts from the bakeries squeezed between butchers and beauty salons. As the night sky darkens, men sell bright, translucent balloons, a symbol of hope in a drowning country. Others exchange Afghani coins for a haircut and shave, trading jokes with the barber like it might be the last time.

It could be the last time. The Taliban mandated long beards in their last reign. Yet we all laugh and smile and dare not utter a word. Nobody wants to remember the last time.

We did not know what the days to come would bring for Afghanistan, but few could have anticipated the Taliban would dominate so quickly, and overnight, plunging daily life into a realm of the antediluvian.

K FORCE
MAND

Within just a few days, the doomsday would swell at a pace we could no longer ignore, we could no longer sanitize, we could no longer stop.

Crammed into old Toyota Corollas, hundreds of Afghans flee their once-sleepy villages as the Taliban wrestles to take back jurisdiction. Some are headed to the homes of relatives in the government-controlled capital of Kabul. Others are lucky to find a vacant patch of hard earth on the sweltering circumference of the city, squished between two roads.

Decrepit tents crop up for the newly displaced, although a mere tent is out-of-reach for the already impoverished. Most are relegated to finding a glint of shade below the decaying, gray tree branches.

They call it a "park," only there are no swings or signs of laughter and life. It's a sprawling plot of arid dirt on an intersection choked with zealous street vendors and beggars begging with glazed eyes.

Little girls smile through their sadness, women weep, and hunger burns through the hollow lives of those who plead for someone, anyone, to listen to their misery.

"I am now a widow. I have no money. I had to take a loan to leave and have nowhere else to go," Halima, a villager from the northern Kunduz Province, cries.

She pulls up her dress sleeves to expose bitter scars from bullet wounds.

"Many bullets have hit me," she says, her voice cracking with emotion. "All we want is for the Taliban to join the peace. We want the peace."

جوس و برگر یاران
Super
سوپر کولا
Cola
0799178707 - 0795950195
GOLD'S
HAPPY
مه برگر

Halima, who is in her forties, explains that her late husband, a police ranger, was killed three years earlier amid fighting against the Taliban. Then two years ago, her two sons were slaughtered in a rocket attack during a Taliban assault. As her only possession and only protection from the sun, she carries their identification documents on top of her head.

"I am here with my daughters. I told them we had no choice. We couldn't bring anything," she stammers between guttural sobs. "We worry about the U.S. withdrawal. We would be happy for them to stay. Maybe then we can go home to our villages."

The pleas are futile.

This is not an official camp with government support, which highlights how brazenly unprepared the U.S.-backed presidential palace is when it comes to provisioning for the final showdown.

As the American drawdown accelerated in late fall, the number of internally displaced persons (IDPs) surged in the struggling nation. However, this parched park filled with misty-eyed mothers and terrified toddlers is just a snapshot of what would follow after the Taliban took over and a horrific humanitarian catastrophe unfolded.

Almost every person you encounter comes armed with documents and desperation, fueled by the perception that any westerner can take their hand to safety. And then there are those scarred by the decades of outside intervention.

There is a shrill, passionate cry emanating from behind the soiled burka of the woman standing at the front of the overcrowded, grassless park. Small children play in the dust near her feet as she lunges forward, her voice rising.

"The foreigners," she bellows, pointing toward us, the few unfamiliar faces, "they cannot be here! They are the reason we are here in the first place!"

Her screams shatter into distraught agony as other veiled women try to calm her down. Having just hours earlier been forced to flee her home in the northern wedge of Afghanistan, she is traumatized and enraged.

One can hardly blame her. Afghanistan has long been pulled apart by war and geography, intruders, and unpredictability.

I move through the mobs of pained faces rippling around me before stopping in front of a father with a leathered face and twitching eyes, absently cradling his young, as if consumed by memories clouded by conflict.

His child, porcelain-skinned and oblivious to the circling chaos, will one day grow up and come to learn the rough hand he had been dealt as an Afghan caught up in the forever war, forced to leave their home and what little they owned.

What would you do if rockets barraged your home and you had nowhere to hide?

What would you take if you only had ten seconds to fill a plastic bag?

Where would you go if every street you turned onto was filled with fighters?

Who would you save if all your children were buried beneath the rubble and crying?

What would you say to those broken and bleeding, widowed and screaming, sleeping on the grime beneath the scorching sunlight, not knowing if those they cherished most were dead or alive?

What would you want someone to say if that was you who had run far from home with nothing but a graveyard of memories?

These are questions most of us will never have to ask ourselves, yet they're questions many Afghans have no alternative but to ask themselves.

Hundreds had already walked with shredded shoes through the mountains or had flagged down an enfeebled cab with their life savings. Yet they survive because human life means something. They survive because there is still hope. There is something better. There is a light. Surely, there is a light: a life where one can sleep soundly and know they won't be awoken by mortars raining down on thinly-thatched roofs.

How willing would you be to rebuild your life over and over again, knowing each time you do, that everything you built with your raw, calloused hands could be taken from you at any moment?

Afghans survive because they have to, and they survive because they want to. Resilience.

How do you measure resilience?

Could we ever measure our own resilience?

Mariam Sayedzada, a local women's rights advocate and camp representative, stresses tearfully that the makeshift displacement camps crowding Kabul are also dangerous places.

"There are no facilities like toilets, there are only a few volunteers, so it is very difficult to manage, and each hour more and more people are coming," she murmurs. "There is no security, and we need to provide a safe place, especially for the ladies."

All most Afghans want is to be safe. But such a space feels like an absolution that may never come.

Two days later, early on a Thursday morning as the first sprinkles of morning light move slowly through the arches of the many mosques, we touch down in the northern city of Mazar-e Sharif, capital of Balkh Province.

The city, raked by heat and grit, was once an ultimate bulwark of anti-Taliban resistance led by Marshal Abdul Rashid Dostum, the expanse to which U.S. forces first entered Afghanistan in the fiery consequence of September 11.

The noose is tightening around Mazar, as the insurgency captures strategic cities and roads all around. But on that bright and dry Thursday, one could never have guessed it was a city being propelled to the threshold of the fall.

Mazar would hold, all the Afghan intelligence officials had insisted. No one would hand the resistance garrison over to the Taliban.

The open-air markets burst with activity. Ice cream sellers shriek with delight as they show off their stirring dance—melting the ice into the cream—in sync with one another. Kebab cafés swell with music and soap operas play on a loop.

It is still normal to sell American-branded goods, and even to drape one's street stall with a "Trump 2020" flag retrieved from goodness knows where.

خوراکه فروشی نجیب
عرضه کننده انواع خوراکه باب و نوشابه های غیر الکولی
0793207080 - 0781414244 - 0797889000 - 0799354555
ارت و میتیو
موجود است
ROSHAN
MTN
خوراکه فروشی نجیب الله
کارت و میتیو موجود است
FLASH

Strangely, there are no Afghan police presiding over the wide streets, and few military convoys roll through. There are no sounds of airpower charging through the skies, even though we know the battles are raging tens of miles away. There is an odd Wild West feeling to the city.

By the following day—Friday, the only day off in Afghanistan, dedicated to morning prayers and family afternoons—something sinister had wafted in.

We head for Mazar's haven in the middle of mania, a calm before what is fast becoming an all but certain storm: the Hazrat Ali Mazar, a mosque which Sunni Muslims believe holds the tomb of Ali ibn Abi Talib, the fourth rightly guided caliph.

Submerged in flowing burkas, women drift in and out of the historic, blue-shaded mosque, entangled in verse and prayer. Faith gives Afghans a sense of affinity, a bond between tribes and strangers, a profound reason to live, and tranquility should they die. Men sit cross-legged in small circles beneath the trees to blot out the scorching summer, children splash in the fountain of fresh water against the backdrop of thousands of fluttering white doves.

On this Friday, they do not fly, appearing more like robotic toys clinging to gravity.

Less than a thirty-minute jagged drive away, Taliban insurgents are combating local and government forces, choking the northern city as the U.S. disengages from its "forever war." Members of the Afghan National Army Commando Corps insist to me, over and over, that Mazar will not fall.

But the civilians, it seems to me, know more.

Halimsah, twenty, arrived in the city years earlier to study economics, after violence had rocked his home in Badakhshan Province.

"There is too much war," he sputters. "I just want to go to the university, and now I don't know."

The youth of Afghanistan know nothing but conflict and the subsequent fragile security bubble brought about by the U.S.-led presence, which has almost popped. Many know nothing about why they are even in this tumultuous situation in the first place.

When I ask Halimsah about the attacks of September 11 and Usama Bin Ladin, he stares at me strangely.

"I don't know," he responds. "Who is that?"

In many ways, it is weariness, worry, and fear from decades of Soviet occupation through the 1980s and Afghanistan's civil war of the 1990s that have heaved open the floodgates to the Taliban.

Many Afghans seem too tired to be angry. Many feel guttered, violated by all sides of the war effort. They can no longer put all their weight behind a crooked U.S.-supported government that does not have their back.

"The president has the power to do something, but he doesn't do anything. Everyone is corrupt in this country," laments Muhammed Lal, twenty-three, who has been forced into the internally displaced camp. "Animals have value in this world. But if someone is Afghan, they don't even have the value of a dog. We are worse than dogs and animals."

Lal examines me with a slight hint of distrust, exhausted by those from the outside just as much as those on the inside who contributed to the staining of his birthplace.

"Is there anyone left who can hear our voice and listen to our pain? Because if you could," he continues, "you would surely lose your mind."

Lal taught himself to speak English and became a high school English teacher.

"I learned it so that I could run away from here," he adds.

I think of a middle-aged shopkeeper, Khalil, whom I had met days earlier. He longs to live anywhere there is peace. Consumed with guilt and sadness, I wonder if Khalil will ever experience such a novelty.

He had turned slowly toward the scorching midday sun and cocked his head.

"Do you think America will see what is happening, and change their mind and stay?" he'd asked.

The body never lies. I rise early that Saturday morning, a strangely sobbing mess. The tears brace me for something I do not yet know.

Jake and I meander slowly through the oddly empty Mazar. The city's dynamism is stripped into shuttered stalls, as hundreds line up outside the banks to withdraw whatever cash they can in a last-ditch bid to head somewhere else.

There is no longer a way out for us. We are booked on the earliest commercial flight back to Kabul, but that is not until Monday evening.

Frightened families are stuffing themselves and their bags of belongings into tiny, intricately colored rickshaws and retreating into the city center, their homes on the periphery under threat. Shops are sealed closed, and cars and construction equipment lie abandoned beneath the scorching sun.

Around midday, Jake and I climb into a random cab and drive toward what we presume to be sandbagged frontlines, and Afghan and resistance forces putting up the fight of their lives. But there are no fights to be seen nor heard.

"I'm scared," the cab driver unexpectedly croaks out.

Our interpreter, Hami, still on his Nokia flip phone on a call to a friend, turns to us in the backseat.

"The Taliban has just broken through the first of the three frontlines," he blurts out with a bizarre smile, either out of secret pleasure or simply to appease my unease. "We are surrounded."

I assuage my discomfort by working, holed up in the hotel room for that long afternoon, stupidly assuring myself that the city will hold until I can fly out in forty-eight hours. Driving is an impossibility; the Taliban has long controlled critical thoroughfares over much of the country.

Jake and I step out into the haunting blackness of deserted streets that evening. We sit nervously at our regular kebab café, inserted down a dozen asymmetrical steps beneath a spice store. For a little while, we pretend it isn't strange that the cords for the television have been yanked from the wall and that no one else is inside.

Before long, the innate anxiety that prickles your skin and shivers your spine becomes too much to tolerate. We forgo chunks of meat and yogurt and hurry across cracked pavements, through the emptiness, toward our towering hotel blocks away.

As I step into the apparent safety of the hotel, pivoting around to take in the surreal scene, I notice a cluster of men on motorcycles in traditional shalwar kameez dress with austere, hairy faces. Storming into the silence.

That is the Taliban.

Celebratory gunfire echoes through the starless night as we watch from the top floor, moving away from the glass walls as a precaution.

It's falling, I text anxiously to a good friend.

An email alerts me that Monday's flight is canceled.

It is fallen, I text again.

There was no fight, no all-out endeavor to hold on.

And just like that, we say a silent goodbye to the Afghanistan we had grown to know and embrace. From the window of my hotel room, we're unsure of how the Taliban will treat foreigners, journalists, a woman. Jake and I set about plotting how we are going to safely exit.

Through the first aberrant night, I curl up in a fetal position on the floor so no one can see me, and I text with another Afghan friend in Kabul: Sama. She is nineteen, studying pharmacy at university with aspirations of becoming a doctor. Sama knew nothing of her country before Americans came in.

I don't know what any of this means, she writes over and over again.

Days of discord drag on, replete with elastic seconds of ambiguity and frightened hotel staff who stay just to serve and shelter their foreign guests. The Taliban foot soldiers are consistently trying to get inside the blast-proof doors at the entrance.

It is unclear what they will do to me, or Afghanistan.

Yet with each passing hour, the chalked streets steadily come back to life; a new life that also feels very antiquated.

A few days into the Taliban rule, more and more of its members swarm the sidewalks. The color-filled clothing that once stippled the scenery has dispersed into fabrics the color of stone. The image of women and children has mostly disappeared from the public sphere. The joys of music and entertainment now belong to the mortuary of memory, left to gather cobwebs.

Given that Kabul abruptly fell the afternoon after Mazar, with President Ghani escaping abroad without a word, Jake and I are keenly aware that any rescue mission back to the capital is off the cards.

The only resolution, we agree, is to talk to the Taliban.

Two Taliban elders drive to the underground basement of our hotel. With a deep breath and a silent prayer to nobody in particular, we climb into the back of an aging Toyota Corolla.

"Welcome," a man with shallow wrinkles and a long, gray-speckled beard says with seemingly genuine enthusiasm. "We welcome you to our country."

We weave in and out of the streets that make up the sand-blown city as the two elders jovially answer my questions about their interpretation of Islamic Law.

"If you steal, you should have your hand chopped off," the Taliban driver choruses nonchalantly. "If you murder, you will be hung. If you cheat, you will be stoned."

Late that afternoon, a small truck clogged with younger Taliban fighters escorts us past the deserted U.S.-issued Humvees to the Uzbekistan border, wielding weapons and taking selfies at the same time.

They attempt to appear menacing with their small beards and roughened expressions, but they strike me as something akin to terrible Hollywood extras, too thin and far from home.

I turn back to take in the beautiful, bleeding Afghanistan. My phone flashes with a new message from Sama, three hundred miles away.

They are taking everything from me, she writes, refusing to refer to the new Taliban governance by name. *But they can't take away my memories*.

BRIDGESTONE
DUELER H/T

PART TWO:

EARLY DAYS OF THE EMIRATE

"You war reporters," one U.S. military contractor states over dinner in Tashkent, Uzbekistan, "you are a different breed."

Indeed, there is something profoundly twisted about plotting to get back into a place where everybody who lives there is trying breathlessly to escape.

Jake and I sit listlessly for days inside the comforts of Afghanistan's neighbors, Uzbekistan and Tajikistan. Then, just as the last American evacuation flight soars up and out of the Kabul skies, we methodically re-stuff our small suitcases and return.

Just because the West leaves a troubled nation behind does not mean that we must do so, too.

There is a small overpass called "Friendship Bridge" that transfers passengers across the river Amu Darya, which divides Uzbekistan from Afghanistan. It takes a few minutes on a small Uzbek military bus and around fifteen minutes on foot to cross, but it is an expedition into a different time and place. On the Afghan side is a border town called Heraitan, and a few sleepy Taliban sit with guns around a rundown hut beside a faded "Welcome to Afghanistan" sign.

Passport control is a paltry building staffed by a lone Taliban fighter lodged in a cramped office overflowing with files. He utters nothing as he skims our passports and glances at our still-valid visas issued by the former Afghanistan government. He handwrites, in curly Pashto, basic details into a notebook before stamping us in with a fresh "Islamic Emirate of Afghanistan" imprint.

به افغانستان خوش آمدید!
Welcome To Afghanistan

The long and snaking trek from the northern border to Kabul begins. It is almost three hundred miles long, and Google Maps predicts it should take around seven hours to drive. However, we all know it will take almost double that, given the state of the roads—evidence of corrupt construction deals of the past (which involved cheap materials butchered by winter erosion) never completed, and a theater of war that felt as though it would never conclude.

Remnants of conflict idle under the September sunshine.

My view from the window consists of mounds of clay dried to dust. I see delicate girls wrapped in brightly colored hijabs bent over in the fiery heat, carrying heavy bundles on their heads like people in a primeval time and place. Burqa-clad beggars sit with their babies beside bombed-out roads, waiting for those passing by to throw a coin or a bottle of water their way.

Evidence of heavy fighting is everywhere—decimated buildings with their masticated contents rotting under the daylight, broken bases turned gray by relentless aerial bombing, and even mosques razed almost to oblivion. The white-and-black Emirate flags fly high into the sunny sky at ruined outposts that belonged to the U.S.-backed Afghan forces just weeks ago. Dozens of wrecked armored personnel carriers, once the staple of the now-defunct Afghan military, fleck the roadsides, their flattened tires sinking into the ground as painfully thin stray dogs seek shelter beside them.

The expedition is rife with magical moments, too. Lush almond trees and small children squealing with happiness in tiny ponds, farmers offering cups of fresh green tea, and homes encompassed by tendrils of mist on the mountain; they capture a real-life vision of the books I read as a child.

"However black a cow is, the milk is always white," our local driver, Gul, quips as we surge on, emphasizing that this is a proverb that they often repeat in Pashto, meaning that there is so much more that unites us than divides us as humans.

The irony is that this is a jaunt across the country that we could not make previously, given the erratic control the Taliban already had over the troubled land. More than twelve hours later and in the dead of night, we reach Kabul: a shell of its former self.

The vibrancy and laughter have given way to lying low and the persistent hum of anxiety.

Soon, what is left of the foreign imprint will fade into oblivion as the Taliban cements its rule in the streets and the Afghan Republic flag is wiped away.

Outside the once-trumpeted U.S. Embassy, built to the tune of $800 million, courtesy of the taxpayer, the Taliban has appointed painters to strip the colored murals from the fortified walls, replacing the vibrancy with its signature white and black.

Just like that.

And after two decades of annihilation and more than $80 billion in American dollars—dished out to train and bolster Afghanistan's security forces—how was it that the Taliban was able to thrash through and capture the country so quickly?

In the end, it came down to tea and talking just as much as bullets and butchery.

As far back as a year before the fall, Taliban operatives were cutting quiet deals with military commanders and government officials, offering large sums of money and prearranging the surrender of villages, towns, cities, and provinces; all not to the knowledge of those sent to the firing line to fight.

Many of the fights were over before they had even begun.

لا اله الا الله

How odd it is to be sitting inside a stuffy ministry with wooden doors and wide staircases, listening to Emirate spokesperson Zabiullah Mujahid speak from behind a walnut table.

For so long, he was the voice behind the veiled insurgency curtain. The voice with no face.

Now, Zabiullah is a relegated ruler out from obscurity, tasked with helping run a confused country of thirty-eight million. He looks confronting but strikes me as a mild man of even temperament who does not raise his voice, who endeavors to wash away the decades of pervasive terror spread by his organization's iron-fisted

insurgency.

Many Afghans tend to call Zabiullah and his cohorts "Talibans," a plural that personifies the concept that each is a student and that the organization is not especially organized, comprised of misfits and religious devotees, dedicated fighters, and ragtag militias from the mountains.

For better or for worse, the Talibans are now forced to come together. That week, they form a hardline, one-dimensional, interim government of Pashtun members, with no women and no ethnic diversity.

Zabiullah's words of defense waft away in a mental blur. I find myself drifting, thinking of something else.

The twentieth anniversary of September 11 is in a few days. What we were doing and what we thought when we heard about those attacks, for those of us old enough to remember, is forever planted in our subconscious.

Twenty years ago, I was a high school student cocooned by books and ballet classes and boyfriends, watching my tiny television in fear thatthe whole world was falling apart. Still, I did not know then what impact Afghanistan—the faraway land on what felt to me like the shores of existence, where the horrible plot was brought to life by al-Qaeda—would come to have on my life, not just personally, but professionally, too.

Yet in the elastic years after the 2001 attacks, after U.S. forces had supplanted the Taliban regime from power, it was hard for me to process how and why, so many expended lives later, that we had returned to the Taliban rule.

"They are bombing al-Qaeda, the Taliban, terrorists, they say," a weeping woman once said to me after losing her husband in an attack on the eastern edges of the country. "But they are bombing my people, too."

Who could ever have imagined that the Taliban would rise again?

Every Afghan has a war story, even though almost all of them never chose to go to war.

There is an extraordinary place poised like a lighthouse in the heart of Kabul, where those with frightened faces and heavy souls wait patiently outside. And inside, shards of light scatter across the neatly trimmed courtyard.

In a fleeting world where organizations are like ships in the night, the EMERGENCY hospital for the war-wounded stays. Emblazoned with red walls, the facility has been there under the tutelage of the late and great surgeon Dr. Gino Strada since the first Taliban rule in 2000.

It is a place as necessary as it is tragic: teeming with broken bones and bullet-battered children. The narrow metal stretchers are occupied with those who have lost limbs, livelihoods, and loved ones in the carnage that has gripped Afghanistan.

But burrowed inside the heartache, places like EMERGENCY are beacons of hope.

A boy with a hollowed chest and protruding cheekbones vacillates between anguish and laughter. His name is Abdul, and he thinks he is about fourteen or fifteen, although EMERGENCY staff predicts his actual age is probably closer to seventeen or eighteen. There is no official birth record, and, likely, nobody will ever really know.

Abdul lost both his legs days earlier to a landmine in Logar Province. Yet he smiles and folds his frail body into prayer, accepting his new reality with a shrug of surrender. When one carries a faith so staunch, and believes so fervently in the power of prayer, one can never be too lost or lonely, or so it seems to me.

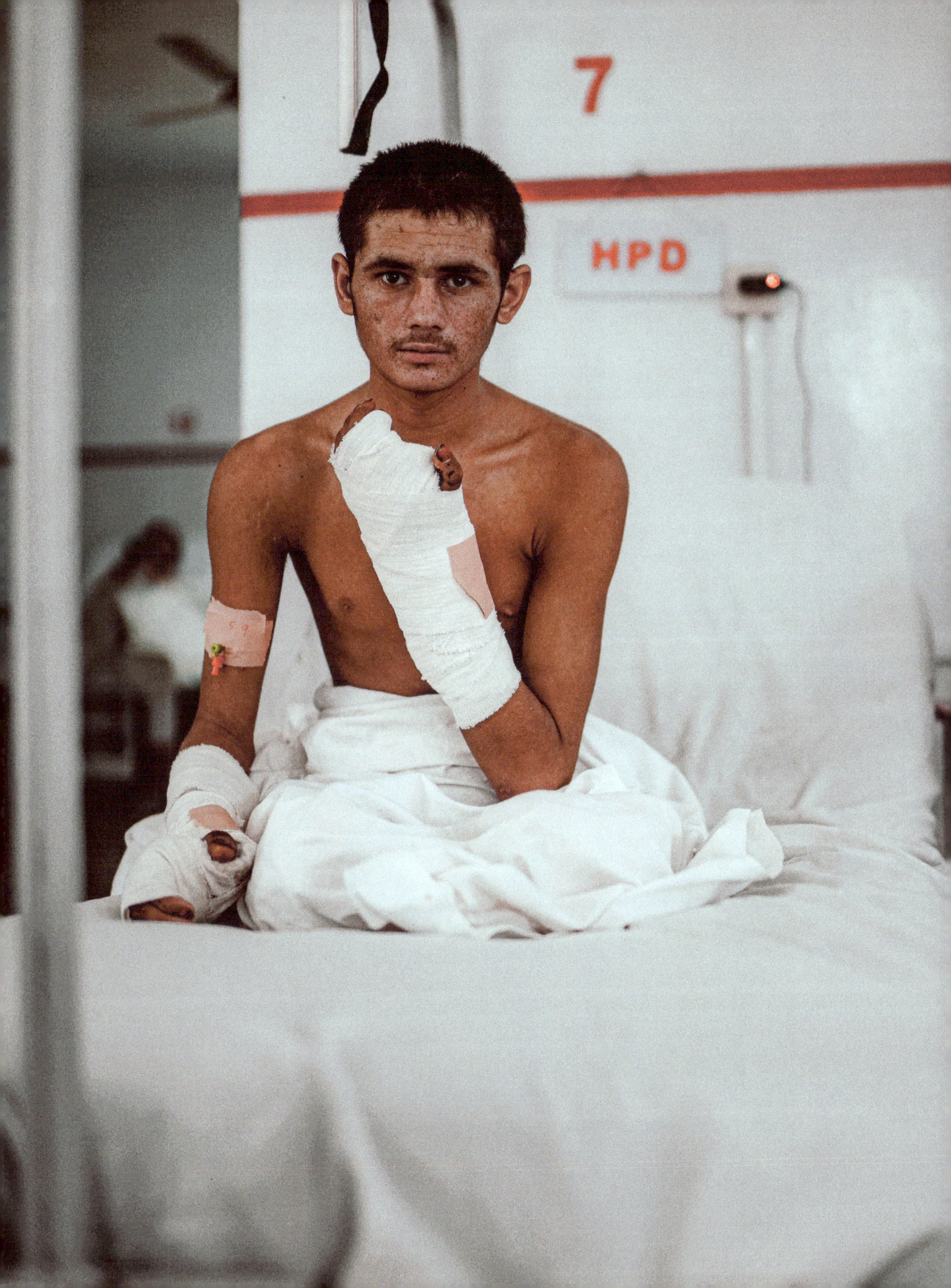
7
HPD

The moment is further fractured by the howl of a small boy on a stretcher, his mangled lips calling out to his mother, yelping like a wounded animal. Nearby, a shrapnel-blinded man reaches out into nothingness—exposing arms stripped down to deep layers of pink.

As we move out into the midday sunshine, the chilling quiet—not to be mistaken for serenity—is ruptured by the loud and ceaseless barrage of bullets being fired on the streets directly outside.

When there is a momentary pause, when you think it has stopped, the battery begins again.

"They are trying to get rid of the protestors outside," one local EMERGENCY staff member exclaims as we shuffle into the center of the manicured courtyard. "It is many women upset, so they are marching to the White House!"

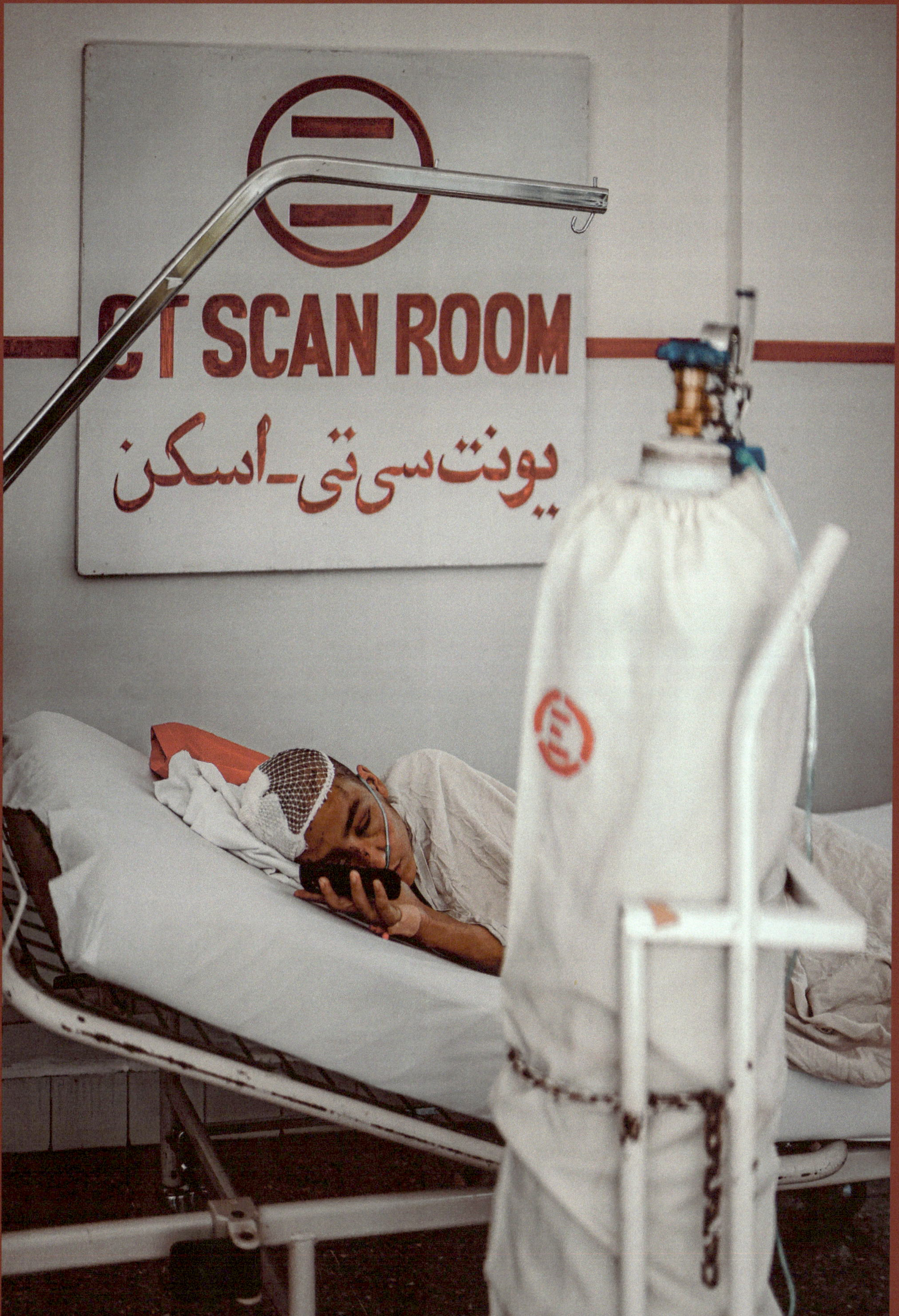
CT SCAN ROOM
یونت سی تی-اسکن

The irony of his words, somewhat twisted in translation but referring to the nearby presidential palace, are not missed. Several feet away, hundreds of youthful Afghans, men and women and small children, carry signs and repeatedly chant the word "freedom."

It has been just over a week since the U.S. left, and the young behave fiercely, as though they have nothing left to lose. Many feel let down by the nation that for two decades urged them to seize equal rights.

مرکز تخصصی و زیبائی
دندان انصاری
Ansari
Speciality
and Cosmetic
Dental Center
موتر فروشی پسران رسول شکوری
مرکز تخصصی
دندان انصاری
Ansari Speciality
Dental Center
Emirates
+93 (0) 799 314 609 | +93 (0) 20 220 1016

افغانستان

Before the momentum could build any further, the Taliban dispersed the crowds by firing bullet after bullet into the still air, prompting scores of demonstrators to run wildly in various directions, disappearing into the throngs of shopkeepers and curious onlookers.

The mass movement is one of several that has unfurled in the transformed city in recent days, with this one spurred by both Afghan women incensed by their loss of opportunity and men riled by the Taliban's incursion into the last remaining province of resistance, Panjshir.

"We're tired," one protester pants, sheltering behind a line of swinging dead cows in a butcher shop, after having run from authorities. "This is the only thing we can do now."

Yet the midday mayhem—awash with arrests, beatings, and camera confiscations—does not deter the women. A few hours later, another protest, filled almost entirely with women in their twenties, proliferates in the same Shahr-e Naw neighborhood. The hundreds of women amass in small packs on street corners before embarking on a daunting walk into a potential hail of gunfire and handcuffs.

"We want what we worked so hard to achieve," an Afghan women's rights activist tells me, unflinching. "We don't care if it costs us our lives. We have no life anymore as this is."

Not only has the newly installed Islamic Emirate of Afghanistan disbanded the Ministry of Women's Affairs, but the government has replaced it with the much-feared Ministry for the Propagation of Virtue and Prevention of Vice.

I am directed to sit underneath an overhanging chinar tree as the Ministry's re-opening ceremony—jammed with the Taliban's top-tiers and Mullahs—wages on across the lawn. As a woman, I am not permitted inside the mosque, although Jake and our local producer, Naweed, are welcomed in to photograph the esteemed affair.

Yet hidden in the homes outside, the Taliban's reinstatement of the Ministry, which was abolished during the U.S. occupation, is sending

a collective shudder through many Afghans, who remember it for its strict execution of Islamic law.

"Our main purpose is to serve Islam. Therefore, it is compulsory to have a Ministry of Vice and Virtue," Mohammad Yousuf, a Ministry representative, later informs me. "Islam has its rules for major sins. For example, killing someone has its own punishment; if it is done intentionally, if the killer knows the person and intentionally kills the person, they will be killed back. If there is a theft, the hand will be cut off. If there is illegal intercourse, the violators will be stoned."

But the Emirate, desperate for international funding and recognition, would hold off on immediately imposing such punishment.

In these early days of September 2021, it remains uncertain whether the Taliban has managed to defeat the last holdout: the picturesque province of Panjshir Valley. Commanded by dedicated mujahideen fighters who had driven the Soviets out more than three decades earlier, it was the one slice of Afghanistan that the Taliban had never managed to control, even during its first tenure in the 1990s.

Panjshir is the coveted acquisition.

Reports flood social media of battles in the far-flung mountains between the National Resistance Front (NRF) and Taliban foot soldiers. Yet confusion prevails as to who is domineering the voluptuous valley once administered by the late Ahmad Shah Massoud, who was slain by al-Qaeda two days before the September 11 attacks.

Although sealed off by checkpoints to the outside, our discerning and relentless minds find ways to smuggle inside.

Dew swirls around the billowy mountains peppered with mud huts, and emerald waters trickle through the narrow, pebble-filled canals. On the surface, Panjshir Valley is a portrait of serenity. Yet one cannot help but feel the ambiguity and tension hidden in the hills.

As we travel through all eight districts untouched by destruction and damage, it is clear that Afghanistan's new ruling power has secured the firm grip of authority.

پیمان
شهید

MASSOUD FOUNDATION

Dozens of robed Taliban fighters congregate outside the Panjshir Revenue Department. They hail from Farah Province some five hundred miles away and belong to a unit rapidly deployed to places as necessary. Taliban commander Mawlawy Khalid claims that all of Panjshir was fully seized three days earlier, yet some Panjshiris had fled to the mountains.

"We will give them a deadline, today or tomorrow or whenever, to surrender," Khalid affirms, adding that whoever does not surrender "will be shot dead."

The process of surrender is that once weapons have been checked and handed over, the individual is then issued "a letter" indicating that they are "free" to return home as usual.

The Panjshir enclave has earned something of a mythological reputation over the years. First, it was the triumphant hub against Soviet occupation in the 1980s, and then the only Afghanistan province not to have fallen to the Taliban during its last 1996-2001 rule. Then, as the Taliban had rapidly captured the country amid the twilight of the U.S. retreat, Panjshir served as the last bastion against the Taliban onslaught.

Almost every shop alongside the main, powdered road that snakes through the province is padlocked closed. And nearly every vehicle—and foot soldier—I see belongs to that of a Taliban unit brought in from provinces far away. The fighters all bear sinewy faces from endless days on the battlefield and almost all prowl with an American-made rifle in hand.

I see spatterings of local Panjshiri men, usually huddled in small groups outside abandoned storefronts. Some jump to their feet at the sight of an armored Taliban truck rolling through, and others wave in a commixture of fear and curiosity.

I do not know what will become of the majestic Panjshir and cannot say for sure if, or when, the fighting will stop. However, I do not see nor hear any skirmishes, except for what sounds like two rockets launching in the distance as the sun begins to set.

Nonetheless, the pain of conflict across their country is wearing thin.

"This has been going on for years," chortles Abdul Hay, who guards a mosque in the provincial capital of Bazarak and swears the religious building was damaged by bullets days earlier. "People are bored of it, and other people are afraid."

Back in Kabul one bright afternoon, we are driven to a nondescript building on the fringes of the city. Stuffed toys wither on a shelf near the building's entrance. Playground equipment remains dead still beside Taliban uniforms draped over the playpen fence to dry; glaring reminders of the nursery school that existed inside the chipped pink walls just ten days earlier.

The former school now serves as the new base of the Taliban's elite special forces unit known as "Badri" or the "Badri Command." But it is not only home to the hardest of hard fighters, who tramp the grounds clad in camouflage and tout an arsenal of American-made weapons; it is the base for those training to become suicide—or martyrdom—bombers, too.

"This Command has two parts," one high-ranking fighter, Hafiz Badry, tells me in a low voice. "There are those who train to be special forces fighters and those who train to be special suicide bombers."

One has to "have done special actions designed for this Command" before being picked for training, Hafiz explains.

He and a fellow top fighter, Qari Omadi Abdullah, twenty-six, concur that they are overwhelmed with those wanting to be chosen for that acclaimed cadre.

"It is not about the picking in this case. It is about the eagerness," Hafiz goes on with enthusiasm. "There are some fighters who even come to us crying and begging, asking why we didn't pick them for the suicide squad."

While the Taliban is now reverberating a message of "peace" and "general amnesty" in the conquered country, suicide squads will remain a bitter staple tactic for the insurgency-turned-government.

Azizi Bank

The Taliban choruses the war is over. However, Afghans are facing a war of a different, just-as-painful kind. It is a war without money, without jobs, without medicine, without support from the foreigners who propped it all up for twenty hard years.

After the Taliban seized power in August 2021, the United States immediately froze $9.4 billion in aid earmarked for the crippled Ashraf Ghani-led government, as did the World Bank and International Monetary Fund (IMF).

That means already-hurting Afghans can hardly access money from their bank accounts. Afghans travel from far and wide to take out a maximum of two hundred dollars per week of their own money. They queue up for days, roasting beneath the September sun in an afflicted plight to feed their families amid an economy trembling toward collapse.

The sounds of music have dimmed into a self-censored silence. We pause, we gaze for far too long at the markets from a broken window where Kabul's finest orchestras weeks ago were still learning to play. We linger, remembering what used to be.

You still spot occasional acts of rebellion: people smoking cigarettes in the street or men wandering in skinny jeans and a fake Western-designer shirt. You still see visceral signs of people clutching onto the past and moving forward at the same time.

"You want to buy a Taliban flag?" a child, who I guess to be around eight with a sun-licked face and light eyes, asks after our taxicab comes to a halt in the gridlocked traffic.

His determined face cracks into an almost cheeky smile.

"Or you want this one?" he continues, pulling out the red, green, and black flag that one month ago still flew overhead the presidential palace.

It is the stuff of nightmares: a car burned into oblivion, another vehicle smashed and singed. Children's toys and shoes blackened and blown apart, windows smashed, and doors lurched from hinges on impact.

U.S. authorities claimed they had thwarted an ISIS-K attack with a targeted drone strike on a terror bomber less than two miles from the Hamid Karzai International Airport (HKIA) in Kabul. It took Pentagon officials nineteen days to admit that all of this had been an error.

Instead, the strike had hit a longtime U.S. aid group worker, Zmaray Ahmadi, thirty-eight, as he had pulled into the driveway of that now fatally-wounded home. The ten victims—three adults and seven children—had all lived together, with fifteen other extended family members, including Zmaray's three brothers.

Zmaray's youngest brother, Romal, had been sitting in the living room preparing sweet tea when the bomb fell. He lost all three of his children in less than a moment: daughter Ayat, two; and sons Bin Yamin, six, and Arwin, seven.

"No one has apologized, no one has helped us," says Romal, half-moons of darkness below his glazed eyes. "[The] Americans can't bring back our brother, our children, our nephews. So if they apologize, that would be sufficient."

Romal speaks softly and calmly, a model of composure. Yet his eyes constantly flit to the floor as if trying to contort his head around the tragedy that took so much of his family.

"Nobody is normal now, including the neighbors and the elders and the kids," he presses on, his throat catching on his words. "If we tell the children, 'Let's go home,' they say no. They remind me that home is not a good place to be."

In the sewage-splattered alley out front, children play; some wear smiles, and others stare straight ahead. Across the way, a woman, Parisa, demonstrates how the blast blew out parts of her windows and cement home, too.

"I buried white brain matter from the incident in my yard. There were body parts on the tree, on the roof, and on the ground. We have washed all the flesh and cleaned up," she says, motioning to the angry purple scars where shrapnel burned her face.

Just weeks earlier, embarking on the infamous land route from the political capital of Kabul to the Taliban's religious capital of Kandahar—thus passing through Logar, Ghazni, and Zabul provinces—would have been a death wish.

For years before the Taliban took full power, the three-hundred-mile expanse was ripe with shootouts and bomb blasts between the then-insurgent outfit and its enemies—namely the Afghan and American forces and their partners. Terrorist and criminal groups have also conducted frequent armed robberies, raids, and kidnappings on courageous, beleaguered journeyers.

Following the U.S. invasion in 2001, the building of the road was described as one of the first major hearts-and-minds missions, with the aim of facilitating trade and ease of access for Afghans. Rather, much of the money was siphoned off into corruption. Afghanistan emerged as a cash cow, littered with construction companies subcontracting out cheap labor and using materials that could not sustain the weight of overloaded trucks. Each vehicle left deep gorges, compounded by heavy rains that caused flooding.

The strategic span was left to crumble into a highway of hell.

The irony is that now the Taliban, the same group that attacked the prized artery for decades, is in charge of repairing much of their own maleficence.

There are remarkably few checkpoints dotting the storied stretch. Most are made up of underage, sun-whipped Taliban fighters seeking shade beneath tethered tents.

At the sight of me, a woman, they avert their eyes to the ground and hurriedly wave our vehicle through.

During the last flashes of daylight jolting through Zabul, golden dust devils whizz through the pink sky, while children emerge from their hard labor, hair and eyelashes black with soot.

We swerve by small, deserted villages, once transformed into battlefields and now left to wither. Later, the flesh-colored fields zooming by give way to tufts of green, and fruit stalls manned by sellers boasting that Kandahari grapes are the sweetest in the universe.

PAMIR

We sputter into Kandahar City: the bustling, orderly flagship where the Taliban was hatched a quarter of a century ago.

Le ciel

Our small car bumps through the narrow road lined with the crinkled faces of children and primitive mud huts that seem to sprout up from the earth. Despite the trillions poured into Afghanistan, the route from Kandahar city to the secluded mosque in the village of Sange Saar—where the Taliban all began—is a scalding reminder that the country is still one of the most impoverished on the map.

Inside a single room with a dank brownish prayer mat sprawled across the cool cement floor and a broken clock on the stained wall, emblazoned with the word "victory" in English, Mullah Mohammad Omar brought together a collection of cash-strapped followers to form the Taliban movement in 1994. It was his rallying cry for jihad that birthed the deep-seated ideology no amount of money nor prowess from the most potent and expertly equipped military on the planet—the United States'—could defeat.

"Due to the war, studies stopped here. But still, people haven't stopped coming to pray," says Mawlawi Hayat Khan, forty-nine, who has filled Omar's shoes as the imam for the past seven years.

I turn toward the pools of sun, my eyes darting across the deadened yard, and see a clay room with small squares chiseled out to serve as windows. Until the 2001 U.S. incursion, that room was a madrassa to educate the up-and-coming generation of mountain militiaman. Although it has not been used in the decades since, locals say Mullah Omar's son, Mohammad Yaqoob—the country's newly appointed Minister of Defense, who they remember milling around as a tiny boy—has professed to build them another.

Haji Abdul Majeed Ustad, a seventy-two-year-old village elder in the mosque, stands firm that people were drawn to Mullah Omar's stringent stance against corruption and believed—still believe—that the movement is against self-seeking profiteers.

"Men and women stood with Mullah Omar because he stood against gunmen in the area at that time. People took hammers, shovels, and guns, whatever they could find, to stand against the looters with him," he recalls, growing louder with every word. "That is how jihad started and ultimately reached Kabul. It was the training and teachings of Mullah Omar that people stood for; he was not a thief or someone who would do corrupt practices. That's why he was very much respected, and why people would continue the fight by his words."

Kandaharis are devout people, emboldened with a deep love for their many exquisite shrines and a robust pride for their provincial palaces. Kandaharis are also known among Afghans to be especially fun-loving, laying out picnics on grassy embankments in the middle of busy roads and roaming every inch of Afghanistan, charged by a passionate curiosity to experience their country.

But when you drive a few hours south to the Afghanistan-Pakistan border town of Spin Boldak, the beauty and the bounty fast falls into an illustration of inherent suffering.

It's a searing scene. Some women are covered by dirty blue-and-white burqas and others have nothing more than a threadbare bed sheet to conceal their drooping frames. Men, women, children languish beneath the burning September sunshine.

Afghans here are sick and dying. Some are pushed in wheelbarrows; the concept of a wheelchair unthinkable to the poor and desperate. Although the ailing are only steps from the crossing into the Balochistan cities of Chaman and Quetta, they are denied the entry that once came freely to them.

Perishable goods rot in piercing heat within trucks backed up for miles upon miles. Embattled merchants, who have pushed through the years of conflict with cross-country imports and exports, unrestrainedly lament that they can no longer afford to put food on the table.

Pakistan's sudden shuttering of the border, citing their own national security concerns and the fact that they have absorbed far more Afghan refugees throughout the decades of conflict than any other country, has incensed the new Taliban border guards.

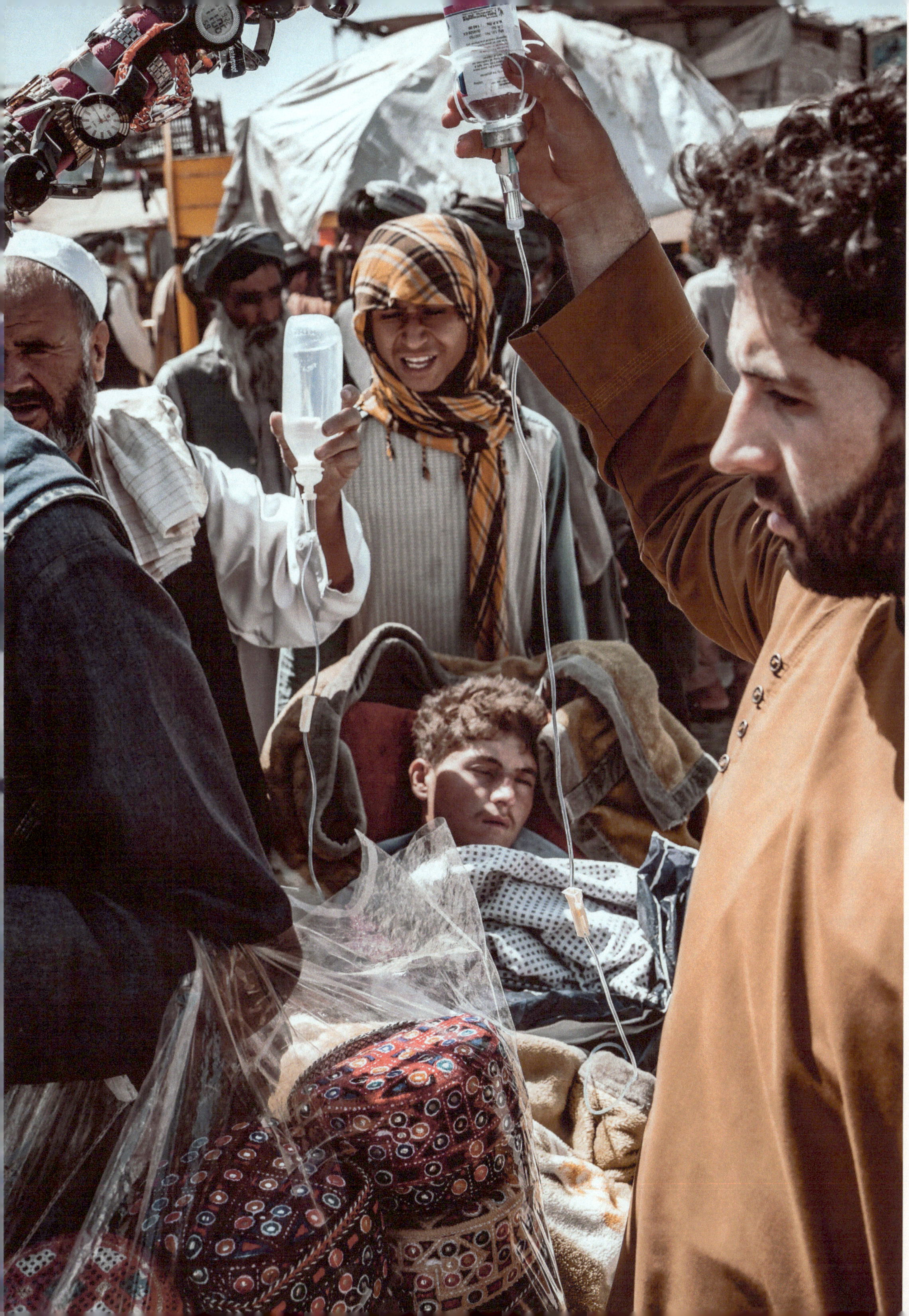

I cannot take a step without being mobbed by Afghans desperate to tell their stories of death and destruction—but is anybody there? Will anybody listen?

The grief embodies a small snippet of a nation sliding into mayhem and madness.

There is an undeniable sense that Afghanistan is teetering, waiting for one hardship to end and another to flare up.

PART THREE:

NEW, OLD WAYS

At the mouth of Afghanistan's second-largest city, inside Kandahar's Central Jail, hundreds of men dot the dilapidated courtyard. Some squat alone outside the heavy cell doors with bloodshot eyes staring absently into the early morning light. Others cluster around a single water fountain, filling plastic buckets to wash away waste in the lavatories, their thin bodies writhing. One inmate prays and prostrates on a large red mat strewn over a cement slab, and dozens wander like crushed shells of human beings.

Without funding for formalized rehabilitation programs, the Taliban has taken to arresting drug addicts from the street and putting them behind bars in a bid to initiate a more orderly rule across the country and eliminate the "haram"—forbidden—practice of drug use.

The dependence cycle often proliferates throughout families: husbands get wives addicted, and then the circuit expands to extended family members and even children.

Inside the Kandahar prison, the youngest addict is just fourteen. His name is Khairullah, and he is kept isolated from the hundreds of adults. The Helmand native appears even younger than his years, with a gaunt, freckled face and stunted frame. He has been incarcerated for about twenty days, having been busted using drugs at a checkpoint.

"I started using four or five years ago because my dad would make me go and bring the opium for him," Khairullah, who only went to school through to the fourth grade, tells me softly. "Every day that I was using, I couldn't even stand up."

Indeed, poverty and crushing conflict in recent years have steadily morphed Afghanistan—responsible for more than ninety percent of the world's heroin, derived from poppy exports—into something of a narco-superpower, reflected by its grapple with rising addiction.

Ironically, the Taliban relied on the illicit trade to fund its burgeoning insurgency. According to one fighter who continues to cultivate in his hometown of Sangin, Helmand, their focus is only on going after those who spread drugs domestically.

The damage it does on an international scale is not their problem.

"We only arrest Afghans dealing to Afghans," he says breezily.

From a distance, the luminous spectacle of white bulbs, occasionally speckled with pink, could seem to be a miraculous display of color, out of place in the parched terrain. However, as you delve deeper into the impoverished farming fields of Sangin, Helmand, the seemingly innocent flowers begin to take on far more sinister meanings. It is not even poppy season in Afghanistan. Nevertheless, some crops still endure across the thirsty pastures.

Mangal, a thirty-six-year-old poppy farmer with blistered hands and leatherlike cheeks, says that some of his crops "had some illnesses" after the U.S. came in, but ultimately his work in the Sangin District was not impacted.

"The Taliban are our leaders," he insists before sipping a tiny cup of tea and pulling apart stretchy, flourless bread. "If they want to eradicate the poppies, we can't do anything about it and will follow."

Janan Mama, sixty, who has reared poppies all his life except for the brief period during the last Taliban rule in which he was ordered to stop, concurs that after the U.S. invasion, he simply resumed.

Children, he explains, usually follow the family lineage into cultivation when they are around eight years old.

U.S. initiatives, led mainly by the Department of State and Drug Enforcement Administration, ranged from dropping poison on crops from the sky, infuriating already struggling communities, to providing financial assistance to farmers if they stopped growing. However, this resulted in even more poppy flowering in an effort to cash in.

These crops have long provided life-saving revenue for the country's remote population. However,, given the broad freeze on foreign aid and the alarming cash shortage, it remains to be seen how the farmers—and the overall Afghan economy—will be impacted by the potential decimation of the narcotics industry.

It is also uncertain if the Taliban has the ability, let alone the will, to genuinely halt the lucrative process.

In the spring of 2017, Afghanistan returned to the headlines. Then-president Donald Trump had ordered the dropping of the almighty, 21,600-pound, guided "Massive Ordnance Air Blast"—better recognized as the "Mother of All Bombs"—(MOAB) on an Islamic State cave composite traversing the Achin district and Spin Ghar mountain range in the eastern province of Nangarhar.

Years on, what became of that heavily debated act?

A shallow crater spans the earth as far as the eye can see. The mammoth hole, forty miles from the provincial capital of Jalalabad, is stuffed with ancient stones and encircled by passels of trees, mud homes, and inquisitive children coated in soil.

Villagers left in the rugged terrain complain of health ramifications. Yet, almost all begrudgingly admit that dozens of ISIS—belonging to the Afghan affiliate entitled ISIS-K or disparagingly-deemed "Daesh"—fighters were slain.

"As soon as the bomb was dropped, all the Daeshs were killed, so people could live in this area. People started coming back. Before that, we couldn't live here. But now there are chemicals here," claims local labor worker Qari, who thinks he is around fifty years old. "Now the people here have problems."

Others say that their children, born less than a year ago, suffer severe digestive ailments, which the doctors attribute to "the chemicals from the bomb."

Locals are quick to pull up their clothing to reveal bright red, irritated streaks across the flesh of their stomachs. Children, who play in and around the carved hunk of dirt, scratch at roughened red bumps oozing on their clay-caked faces. It is impossible to verify the cause of their distress; nonetheless, locals repeatedly contend that the health problems only began in the bomb's aftermath.

What is worse is that they can no longer grow enough crops to make even a basic living, stressing that their irrigation systems no longer function properly.

However, the unspoken woe now is that it may have been all for nothing. ISIS-K is growing once again, launching almost daily attacks across Afghanistan.

Many such loyalists hide in the hilltops and at the foot of the Korengal Valley, furrowed deep inside the Salafist stronghold of Kunar Province.

For more reasons than one, Korengal has been apocalyptically termed "The Valley of Death."

To reach the arresting valley lassoed by towering dunes and thick, opulent forests, one must sputter for hours from the Kunar provincial capital of Asadabad through broken, rock-coated mud tracks, and wind through wild terrain set upon a mountain ridge. Then, finally, the relief of a sturdy yet deserted military base will seep into view, as if trapping everything inside a thirteenth-century time capsule.

The breath-stealing Korengal Valley, which claimed so many American lives, idles beneath wells of sunlight. Seven years after the U.S. footprint disappeared, there is little left to show for the blood, sweat, and tears.

The Taliban avows to have complete control of the Valley and the support of every household, but sentiment on the ground tells a different story. It has morphed into a unique place where almost every family comprises of some Taliban fighters, some former Afghan National Army (ANA) soldiers, and even some who have pledged allegiance to ISIS-K.

"Here, if the father was a Talib, then the son was a soldier in the ANA. Then the brother and uncle would be Daesh. But, somehow, everyone is a splinter group of the Taliban," explains one thirty-three-year-old Korengal local who requests only to be identified as Safi. "That is how it is here."

After the U.S. withdrew from the Korengal in 2010, the inaccessibility of the craggy topography paved the way for outside insurgencies to come and go. As a result, the Valley's villagers all have stories to tell about the six months they spent entrapped by the ISIS occupation around two years earlier.

"It was really hard. The fighting was always there. The robberies were always there. They forced people to go out of their homes, and any wife of a Taliban was made to leave the area," one village elder, a fifty-five-year-old named Mohammad, recalls with a wince after every word.

However, in addition to the mosaic of allegiances even within a family unit, the overgrown, wiry valley bears its own starkly independent spirit—an unflappable determination that nobody but their own tribal elders will lay down the law of the land.

Mohammad exclaims, without boasting, that the Korengal people "got together to beat them," which is how the Daesh eventually left.

On another road trip, we speed along one of the few serpentine stretches of road that has remained paved and intact, almost untouched by the two-decade war that crippled most of Afghanistan's land arteries.

The road from Kabul to Khost—one hundred and fifty miles by way of Logar and Paktika—is a serene journey through the plush and primitive, tucked inside the bloodstained panorama. Yet, the province of six hundred and fifty thousand people comes with a checkered past. It's home to much of the U.S.-designated terrorist organization, the Haqqani network. This thoroughfare was long used by operatives for the group's close allies, al Qaeda, to trundle in and out of the Pakistani tribal region of Waziristan, where extremist madrassas and training camps boom.

Uniquely of anywhere I have been in Afghanistan, here the red, black, and green flag of the toppled Afghan government—the emblem of the Islamic Republic, defeated by the Islamic Emirate—still flies high.

"We won't let them touch our flag," one women's rights activist and schoolteacher says, with squared shoulders. "These Taliban men need to show us they have changed, that they aren't like they were and that they will give women full rights."

Certainly, Khosti women are tough. Many activists still show up to work each day—including teachers of now paused girls' high schools—in defiance of the Taliban restrictions.

Unlike most places that I have visited, the local population does not hesitate or fear speaking out against the new government.

"I remember all the things they did to us in the previous rule, like whipping us and not letting us learn," one grandmother remembers. "They cannot expect quick forgiveness."

Other Khosti friends, like my beautiful friend Nadima, a fashion designer and comedian who runs an Afghan non-governmental organization, DreamVoiceAct, are far more open to giving the new Emirate a chance.

"Who are you?" a Taliban operative asks suspiciously, as she twirls in her lovely handcrafted clothes, while Jake snaps her picture cast against the mist-laden sunrise.

"I am yours," Nadima responds calmly. "I am Afghan."

As the late fall freeze sweeps into the crying country, we adventure for days along the bumpy, broken, unpaved roads, and then on foot inside the wanton and wild valleys of Afghanistan's isolated and least populated province, Nuristan. Young locals, their eyelashes caked in debris, blast dynamite through a cave wall in a desperate bid to unearth treasures.

The men—just sixteen and twenty-three—have placed thin mattresses inside their nebulous cave. They work all hours of the night in the quest to discover enough jewels to feed their families. Others will find wealth with their discoveries, but

They escort us down the mountain and across the valley, to the one-room hut home they share with several other miners. The men excitedly invite us inside for tea and pour us bowls of stale popcorn, an endearing nod to the pious emphasis Afghans place on hospitality. But as we talk and talk, it becomes apparent that their lives, day and night, revolve around scouring the untapped field for enough gems—from tourmaline, kunzite, aquamarine, spodumene, and beryl—to sell at the local markets. If they are lucky, they will make just enough for ends to meet.

It is a deeply dangerous vocation. The men share stories about friends "dying or becoming disabled" due to blasting gone wrong, or entire mines collapsing.

The United States Department of Defense has estimated that Afghanistan rests upon more than $1 trillion in natural resources—from precious and semiprecious stones to metals like gold, copper, iron, and lithium, to ore and hydrocarbons—all ripe for large-scale mining.

With the U.S. now departed from the region and the Taliban in charge, countries hostile to American interests can take advantage. As a result of the U.S. and other global monetary bodies freezing their aid in September of 2021, the Taliban announced their government would be primarily dependent on China for financial assistance.

Perhaps the biggest prize of all is Mes Aynak, planted on a secluded blanket of land inside Logar Province. In 2008, the state-owned Chinese mining company, Metallurgical Group Corporation (MCC), beat out scores of tenders to secure a $2.83 billion, thirty-year lease of the copper-swathed mine.

Back then, and even in 2001, when Washington first entered the fighting fray, there were no smartphones and electric cars reliant on high-capacity batteries and high-tech chips derived from copper and lithium deposits. Cut to 2021, and much of the globe is dependent upon such commodities to function. The Chinese investment—which never fully got off the ground under the previous Afghan government—now stands to be at the center of the world's economy.

Reaching the isolated Mes Aynak requires hours of driving from Kabul through narrow, rutted tracks adorned by white bed sheets used as Taliban flags, blown-out former military bases, and numerous checkpoints operated by bored fighters.

Then it emerges. Mes Aynak drapes atop a sunbaked landscape, beneath an umbrella of blinding light. That stark light illuminates the archaic firepits that were once used to melt metal into coins, the chiseled squares of sun-hardened clay that were once bunkers where townsfolk slept, and the piles of dirt that once were the finest temples belonging to the pre-Islamic era.

Although it means "little copper mine," Mes Aynak hosts the planet's most extensive, unexploited copper deposits, to the value of 11.08 million tons. It is also home to Central Asia's most critical nuggets of archaeology.

The Aynak is a colossal complex yet to be completely excavated. It is ripe with dozens of identified landmarks hailing from the mysterious Bronze Age: Buddhist monasteries dating back to the fifth century, and hundreds of Buddha statues and stupas for devotion.

I imagine the life that once thrived here when the Aynak was at its zenith during the fifth to seventh centuries. I close my eyes and envision hundreds of Buddhists and Hindus entranced in prayer on one side, and the original Zoroastrians—from one of the oldest monotheistic religions—emerging from a blazing fire temple on the other.

The Taliban may soon have to choose between preserving the religious, historic artifacts still untapped far below or destroying them in exchange for reaping the financial benefits from the copper deposits. In a time of dire financial need, the latter hangs in the air.

What this means for Afghanistan or more broadly, world history, belongs to the thick file of the unknown. The Taliban promise to safeguard the nation's time-treasured sites. Yet it wasn't until March 2001, more than five years into their first reign, that the group's leader Mullah Mohammed Omar ordered the famed sixth-century Buddha statues in Bamiyan to be blown up.

Only time can tell.

As the sun dips behind the towering, scarlet rocks of the northern province of Bamiyan, smearing pink across the frozen sky, dozens of Taliban fighters pose for selfies. They're milling in front of what is left of the Buddhas of Bamiyan, which the Taliban blistered and blustered more than twenty years ago, much to the horror of the international community.

The irony.

Now, the former insurgency vow that they have no intention to erase history but rather seek to protect the deep and abundant past entrenched in Afghanistan. The Taliban is on a colorful PR quest; they say they are determined to build the tourism sector as an avenue to halt the spiraling humanitarian crisis.

There is a method wrapped in the madness. Beneath the layers of bruises and bloodletting that have drenched Afghanistan for decades, the nation remains one of the most enchanting on the planet.

4

Taliban fighters float aimlessly on swan-shaped boats across Bamiyan's jarringly blue lakes. Women work from dusk to dawn gathering food and wood before trekking slowly toward their caramel-colored abodes in the distance. It's a raw glimpse of real life, so intriguing it could belong on a big screen.

ADC

مسجد شریف جامع
نبوی

"We are establishing our Facebook page where we will promote tourism and tell people how beautiful Afghanistan is, showcasing the mountains, rivers, valleys, and everything," the newly appointed, thirty-eight-year-old tourism director Mohammad Zubair Musafer states from his Kabul office. "We are going to promote our country. So come and take a firsthand look."

Certainly, Afghanistan is far more than a swath of dirt defined by bombs and bullets. Long before the country was characterized by conflict, it stood at the focal point of great civilizations, imbued by a dazzling medley of arts, languages, gastronomy, and cultural and ethnic traditions.

Breathtaking vestiges are still there, a sight for inquisitive eyes.

From the old-city trading hubs grooved into the traffic-clogged Kabul, and the remote gemstone-peppered caves and valleys of Nuristan, and the Herat Citadel built by Alexander the Great, to the primordial huts at the foot of the Hindu Kush in Panjshir, Afghanistan is a trove of treasures sadly unseen by many Afghans themselves amid intractable conflict.

For example, Ghazni Province, headed by a rundown capital city of the same name ninety miles from Kabul, was once the capital of the mighty Ghaznavid empire that reigned from 977 to 1186 and expanded from India to Persia. The province has long been a Taliban mainstay, rendering it too dangerous for most Afghans to have previously dawdled in for too long.

The Taliban fighters now insist that they are ordered to preserve and promote tourists visiting the star-shaped minarets and mausoleums, a throwback to Ghazni's glory days.

And on the bronze-dirt, quiet hems of Kabul, the Taliban guards pledge that they will soon pick up the pieces in bringing in visitors to the primed Tajbeg Palace—dubbed the Queen's Palace, built for the Afghan royal family in the 1920s.

The elegant, hilltop dwelling bathed in pure white light became the headquarters for the Soviet 40th Army in 1979 and sustained severe damage in the decades-long war. Copious monies were poured into its restoration by the last Islamic Republic government. It had been slated for an official opening on August 19, 2021—just four days after the capital would fall to Taliban hands.

On Fridays, Taliban soldiers gather in clusters to freely explore Zazai Park, a family amusement center on Kabul's periphery, delineated by Ferris wheels and swings and glitter and sugary foods.

They stare, puzzled by the picture of a woman daring to go on a ride. I wave at the kohl-rimmed eyes below in stubborn refusal to be sidelined.

ZAZAI FAST FOOD

Afghan families—including Taliban commanders with their wives and children in tow—picnic beside the fairytale lakes of Sarobi. The expansive body of blue is situated in Kabul District yet feels so far from the overwhelming hustle and bustle.

If you are fortuitous enough to go when the world is wandering elsewhere, you can sit with the everlasting streams all to yourself.

I think of an Afghan driver I once had, a medical student who spoke whimsically of the homeland he loved so dearly but so badly wanted to leave.

"We Afghans are unlucky people," the driver had whispered. "But we would be the luckiest people if the wars ever left—look outside at this magical place."

PART FOUR:

AFGHANISTAN FOREVER

From the outset, the Chaparhar District of Afghanistan's eastern Nangarhar Province looks like nothing out of the ordinary: markets crawling with livestock and fresh fruits for sale, children hawking sugarcane and trinkets, weathered faces peddling rusty bicycles and packed into brightly hand-painted rickshaws.

But drive a little deeper, where there is little sign of the white Islamic Emirate of Afghanistan flag flapping in the breeze, and therein lies a sense that anything can happen at any time. After all, Chaparhar has long been the territorial and operational bastion for ISIS-K, or Daesh.

"There have been more than twenty dead bodies here since the Taliban came," one villager stutters to me from outside the gas station. "Just two days ago, a person was shot dead; we don't know by who."

Down a nearby track, where men dig into the dirt, and covered women roam with babies in their arms, villagers say that the slaughtered number is actually more than forty-five. Several anxiously assert that a man is "taken out of his house and killed almost every day."

Another group of Chaparhar locals solemnly divulge that two days before our visit, a "body was found near their house."

With a population of around fifty-five thousand, the district serves as the base for the terrorist organization to convene and construct merciless attacks nationwide.

Villagers, in hushed tones, spell out that there is a cemetery just past the animated Ghurfa Bazaar that marks the start of the ISIS-K dominance.

"ISIS has been here in Chaparhar since its beginning, since 2015," says one local. "But the Taliban is only here during the day; at night, they gather at a base and don't go out. Even the previous government soldiers would only go out in the day. So this area has always been ISIS."

That graveyard is defined by decayed Afghan flags striped green, red, and black, representing the old Ashraf Ghani-led government. Alongside them are shreds of the signature Taliban flag. Then there are others—ominous, plain black.

"ISIS are there, a few ISIS," the Chaparhar district governor, whose name is Ainudin but is "famous by Badrudin," reluctantly admits from outside his dim compounds on the edge of the Bazaar. "They might be hidden somewhere, but the Taliban has taken over, and we are fully in control."

The farther you accelerate along the narrow Chaparhar Road away from Jalalabad, away from the market and the graveyard, the more apparent the tingles become in your spine. The Taliban checkpoints die out, and there is next to no traffic. Several stores have black cloth covering their windows, while others display a black flag printed with a white Shahada, which looks chillingly like the infamous flag of ISIS stamped with the Islamic oath. If the devotional design is a coincidence or intended for ISIS allegiance, there is no way to tell.

"ISIS acts as an insurgency; they stay off the main road. But if they have a specific target in mind, they come out of their houses and target that person," one driver explains. "That is how they operate."

There is a sense that nobody knows where anybody belongs. And if they do, they are too scared to say.

SBK

Outside one general store, embellished by that jarring black cloth, a teen employee whispers into the phone asking whether the "people coming" are in a car. I remember back to a little earlier in the day, at the market, when a small boy peered through the window at the camera hoisted on my photographer's lap and raced off to tell an adult observing nearby—who immediately pulled out a flip phone.

It is not a place to linger too long.

Since ISIS-K's inception in Afghanistan in 2015, as the outfit gained momentum in its flagship of Iraq and Syria, loyalists have swarmed into the otherwise quaint canton. It is from Chaparhar that ISIS-K has plotted, planned, and dispatched its assailants—from suicide bombers and shooters to those who ambush security checkpoints and plant bombs—to steal the lives of hundreds, if not thousands.

Running afoul to the official Taliban position that the Daesh in Afghanistan is a "small problem" or simply doesn't exist at all, some Kabul-based, Taliban intel heads quietly consider the district "one hundred percent" ISIS-K domineered. It is essentially a no-go zone for their foot soldiers.

"It is the barbaric duties they do—decapitating or removing eyes and things. That activity only gets them stronger," one local utters. "That is how much fear the Talibs have of ISIS."

In the district next door, less than ten miles west of Jalalabad, sits sleepy Surkh-Rōd. Comprised of dehydrated tracks and a daub of pottery-like homes, men and burqa-clad women labor from first light until last. Just after the autumn sun sets each day, residents—and the Taliban themselves—vanish into the night haze.

Ever since the Taliban took power, Surkh-Rōd has morphed into a red zone. ISIS-K has steadily increased its savage footprint—spreading from the neighboring Chaparhar to cement nightly control.

Samiullah, twenty-three, who has lived his entire life in Surkh-Rōd and works as a night guard for a neighbor's home, points out in a barely-audible whisper that "there has been an increasing number of Daesh coming into their village" at the first hints of dusk.

"Generally, it is okay to walk around during the day, but Daesh take people out during the night," he says in a terrified stammer, his face ashen. "They aren't from our village."

Samiullah says that two weeks earlier, ISIS-K operatives took two people "out from their homes," and their bodies soon surfaced—shot in the head with their eyes gouged out—in the village's sewage ditch.

Several other locals corroborate that finding, pointing us to the sinister, parched canal.

"Many killings happen on the street; the Daesh and the Taliban kill each other," explains resident Muhibullah, twenty-eight. "There is dried sewerage, where too many decapitated heads have been found, right next to the main street of the village. The most recent beheading of two people was two days ago."

Samiullah stresses that the Daesh tell people they are killing Taliban, but he is sure that those two found dead in the sewage were "common people," aged around eighteen and twenty-one.

Toward the end of the twisty track sits a comparatively nice, simple brick house furbished with a bright aqua door. In late August 2021, it was hit by a U.S. drone, days after ISIS-K slaughtered thirteen American troops—and more than one hundred and eighty Afghans—in coordinated attacks on the periphery of Hamid Karzai International Airport as the frantic U.S evacuation effort was taking place, days after the Taliban took control.

Inside the house with the bright aqua door lived the alleged evil who plotted and planned those attacks.

The target—or targets—were seemingly struck sitting in a chair outside, with very little damage to the rest of the property. They were identified by U.S. intelligence as the masterminds of the August 26, 2021 airport onslaught.

"The Daesh came to that house after the strike and brought things from the underground. They took out big boxes of small bullets and RPG bullets, and three bombs," Samiullah recalls. "They returned the other day to that same house and took handfuls of SIM cards and some prosthetics—artificial hands and legs."

He also notes that ISIS members themselves came to the suspect brick abode the morning before the drone strike and returned the morning after to "collect the body parts of the dead." But they left with more than that, he utters with a quiver, emphasizing that the vehicle appeared "super heavy," with its suspension crashed in.

"It didn't seem like only body parts in the car with that amount of weight," Samiullah explains fearfully. "They took out something else."

He also professes that three Taliban fighters went to the perilous home the afternoon after the strike, emerging with what appeared to be ISIS memorabilia. Three headless bodies were found hanging from a nearby tree a few days later, along with an ISIS-embossed warning letter.

U.S. officials have cautioned that ISIS-K is a "serious threat" and put forward the possibility that the group could strike the homeland within months, as it purports to replenish its capabilities and restore the gains lost at the hand of the American military and its partners.

But let us not forget that those who have to endure the brunt of the carnage are the Afghans. After decades of ceaseless wars, the weary and war-trodden are not convinced that the fighting has passed. The ISIS violence has provoked many nights of sleepless worry that the country could descend into another war, this time between the Taliban and warring parties.

Or that they will be the next bland statistic sketched into the Afghan handbook of death.

"You should turn around. It's not safe to go further down this road," one Surkh-Rōd villager warns, glancing up from his position digging trenches deep into the earth, his words imbued with a kind of sadness. "We don't know who is there."

A baby's yelps shatter the silence inside the ICU, wedged inside the Irene Salimi Children's Hospital in the cold nucleus of Kabul. The little boy, with a tiny body and wincingly swollen head, calls out in a croaked cry, veins protruding from his scalp and long, thick eyelashes wet with tears.

It is impossible to decipher the infant's age. I soon learn he is eight months old, and his name is Mohammad. His mother, cheeks blushed and bearing the weight of the world on her shoulders, looks on in helplessness.

"He has too much water in the head; it's a congenital anomaly," Muhibullah Ahmadzai, medical director of the Germany-sponsored hospital, says in halting English. "It's fairly common in Afghanistan because most women do not fully eat."

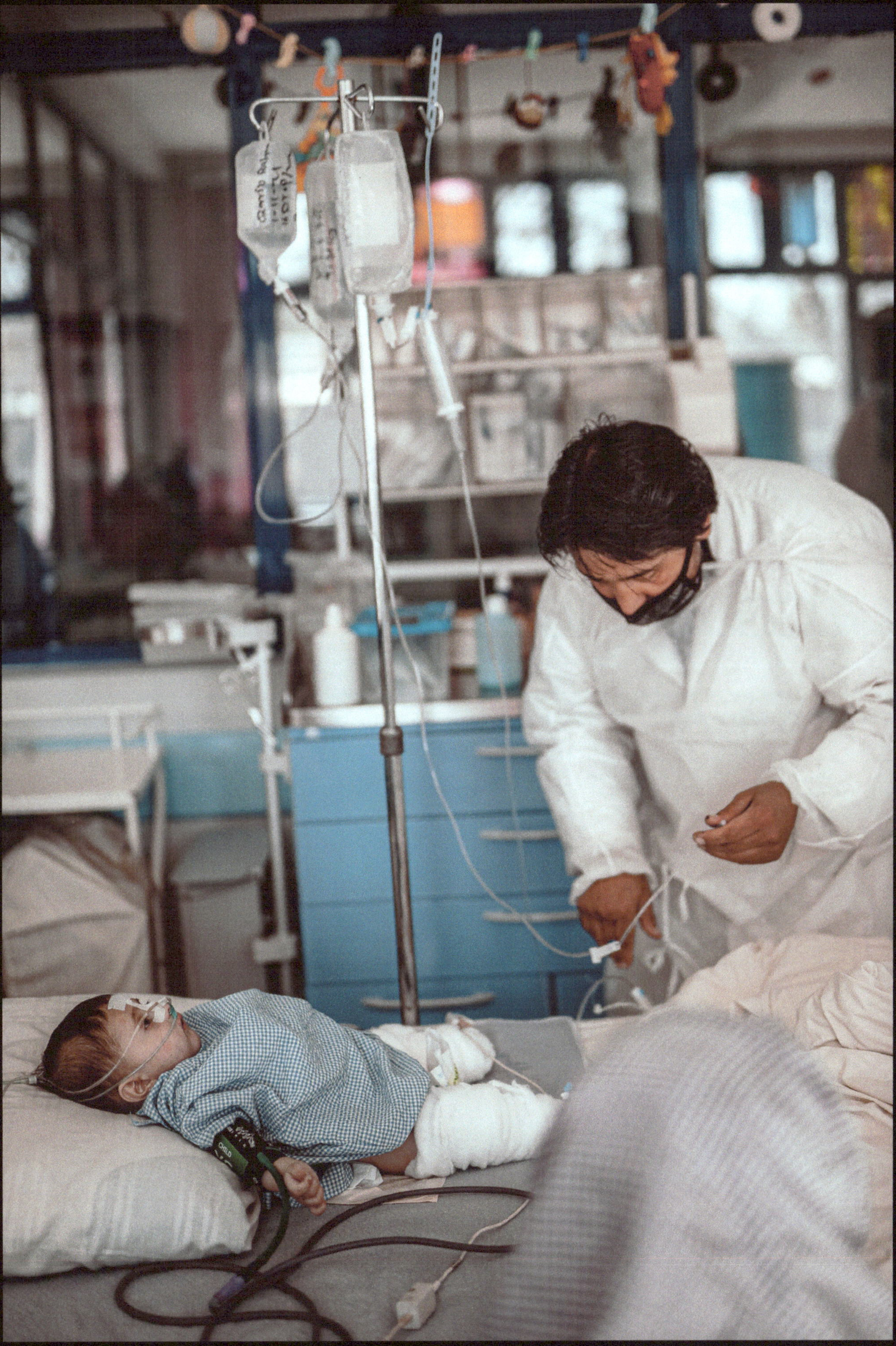

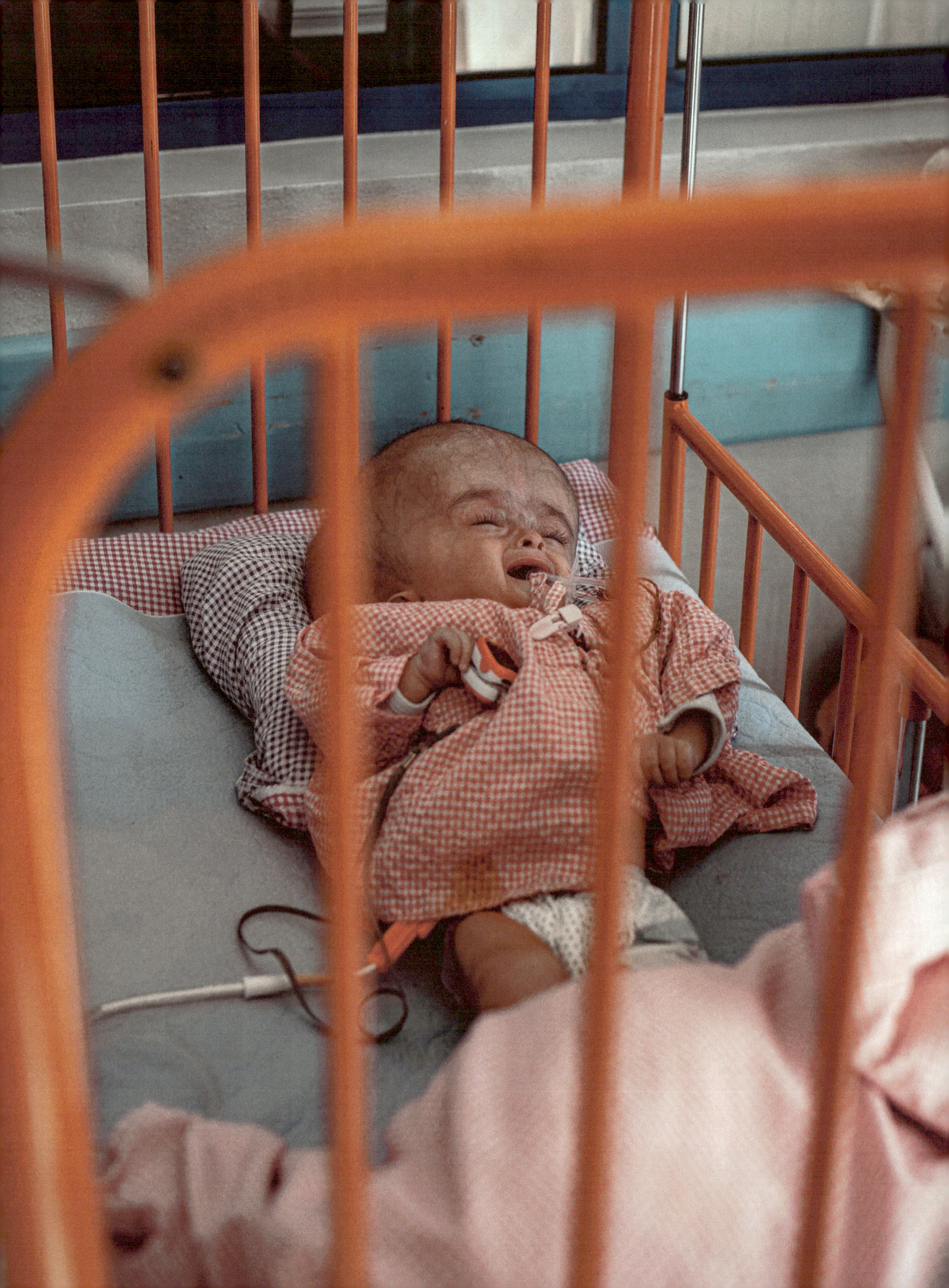

The congenital disability is hydrocephalus, which is characterized by an oversupply of cerebrospinal fluid in the brain, resulting in extreme pressure and pain. In the U.S., one in five hundred babies is born with such a condition. But in war-ravaged Afghanistan, hydrocephalus has become far too familiar amid decades of poverty and conflict.

"During their pregnancy, it is hard for many mothers to supplement with enough iron, so they suffer from extreme deficiency and anemia," Ahmadzai continues wearily. "For six months in the womb, the babies' brains are under pressure, and then even when the mothers give birth, sometimes they cannot seek the medical help they need right away."

Irene Salimi Children's Hospital is a rare respite in a nation in crisis. The facility, sifted with baby blue walls and a lush garden outside, is where the poorest of families from faraway places come for lifesaving pediatric surgery, traumatology, treatment for severe infections, and rare congenital anomalies.

They appear too destitute, too apolitical to care about anything but living another day. Irene Salimi remains a scrap of dignity, a glimmer of hope in a murky landscape.

Swaddled in thin blankets, Nafisa struggles to open her fluttering eyes. She is only five months old but bears the air of an old woman, her pink skin cracked and stretched around her painfully gaunt frame.

"She has lost her appetite. She lacks blood," Nafisa's fidgeting mother tells me, gently, from beside her daughter's cradle.

Doctor Ahmadullah assesses the girl's chart and pulls his shoulders back, projecting an upbeat expression in a job that is rife with darkness.

"Nafisa has swelling and edema, her stomach is upset, and her muscle is wasting," he says stoically. "She is five months old, but she looks a year old; children with malnutrition age very fast."

Baby Nafisa is hardly alone in her fight for survival inside the rundown walls of Kabul's Indira Gandhi Children's Hospital. Three wards of the state-run facility overflow with just the malnourished. Given the lack of room, every two or three infants share a cot or incubator bed, and overburdened, helpless mothers bravely breastfeed on the floor.

For a hospital filled with sick children, there is little sound. Most are too sick, too fatigued, too close to death, to cry.

"There is no medicine. There is no food. There is no salary," Dr. Noorulhaq Yousufzai, associate professor of pediatrics, who has worked at the hospital for twenty-eight years, laments from an unheated basement office. "We have three hundred and sixty beds, but we have more than five hundred patients a day. And if we do not have the right medicine left, a child will die."

Meanwhile, stick, discolored limbs jut from eighteen-month-old Mariam's diminutive dress as her Aunt Aminia watches over. Blonde wisps fall over Mariam's rumpled lashes as she fights to open expressionless eyes.

"She has wasted blood and is so malnourished that her hair has lost its color," Dr. Ahmadullah asserts. "Causes of malnutrition include the mother, who doesn't have good health and lacks the necessary vitamins. This is due to low income, difficulty obtaining food, and various physical and mental health conditions."

The multitude of healthcare facilities once operated by international donors have receded since the regime change, putting even more onus on the shattered public system. In addition, life-saving medical supplies have all but been used up, with little to nothing coming in.

The budding and innocent always bear the brunt of war, in its height and in its aftermath.

As well as the financial catastrophe and the Taliban's status as a pariah government in the eyes of the international community, decades of conflict and drought have displaced millions and led many to the verge of famine.

Winter has now settled across the gaping land.

"We don't have a salary, and that is the situation," another doctor tells me, pain and tenacity sewn into his watery eyes. "But we continue to do this because it is our responsibility. We should save the lives of the people. This is the responsibility of every doctor, and we will fight because this is our country. These are our people."

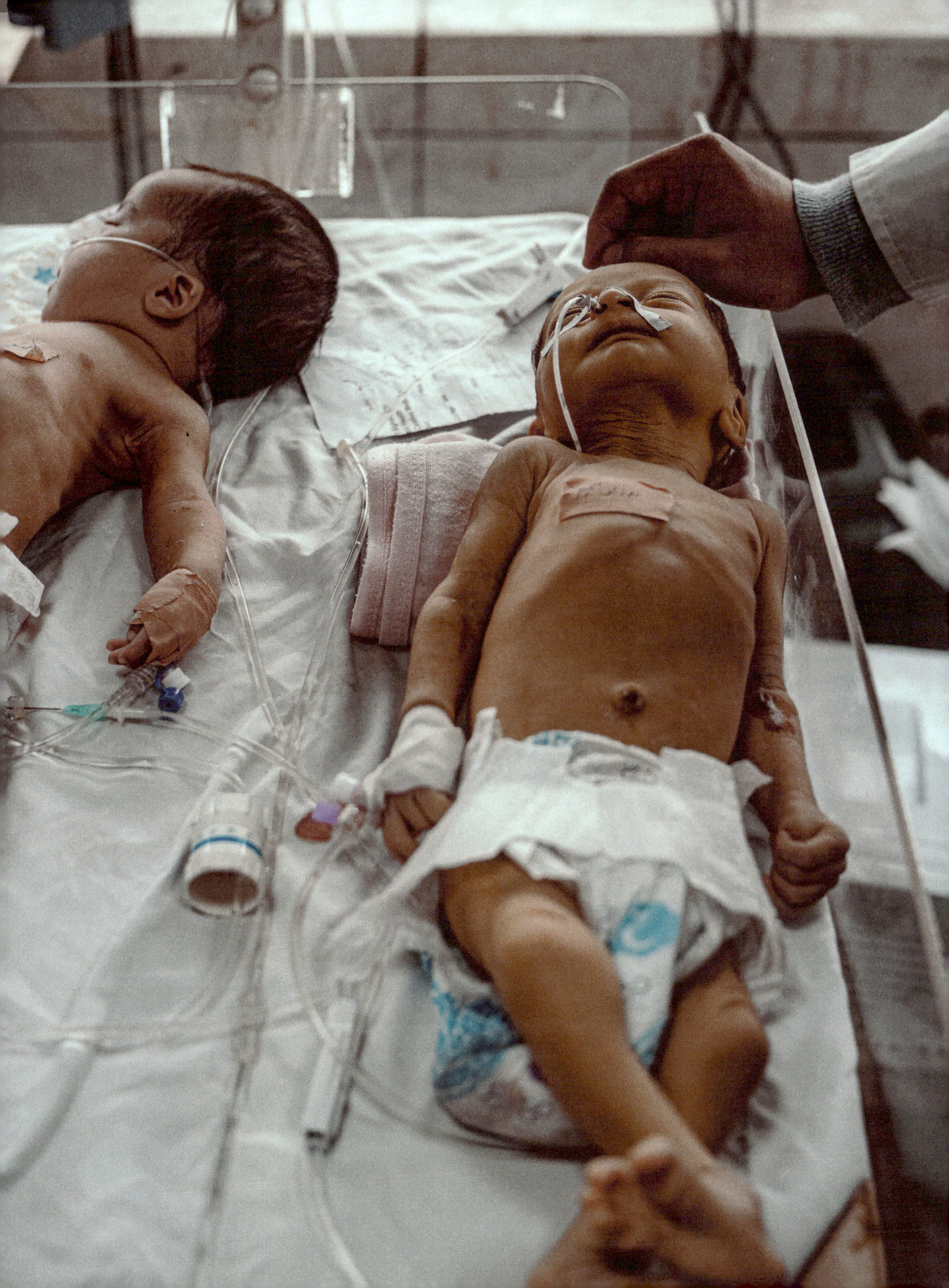

Amputation
Patient Files

The doom and gloom of a humanitarian catastrophe cling as the icy winds roll in and snow glazes the serrated mountains of the beautiful, bleeding country. Unlike winters past, most life-saving humanitarian aid organizations are no longer present. For the few that remain, the scenario is akin to calamity.

"Many people would come to us asking for food, for help, in the past. Now there are twice as many of them. In Afghanistan, everybody is a victim of war," notes Alberto Cairo, the longtime head of the International Committee of the Red Cross's (ICRC) orthopedic program in Afghanistan. "The situation is deteriorating, medical professionals are without salaries, and teachers are at home now. People are poor; they don't have food."

Alberto underwent his first ICRC mission to Afghanistan in 1990 and refused to ever leave. He fell for this land, the way so many of us from the outside do, and his every patient became like a surrogate child. Under Alberto's bubbly, warm direction, countless Afghans caught in the conflict crossfire have had the chance to walk again.

Further north in Herat, where South Asia and the Middle East converge at the Iranian border, signs of unrest—the search for safety—still embattle the Shi'a Hazara minority. In the sand-blasted city and its surrounding villages, too many Hazaras live in poverty. A great many people speak with wispy voices, worn out by years of mistrust and repression, and their voices do not ring with emotion. Many Hazaras save every penny with depleted faces and restless hands, hoping to escape to Iran, which has a Shi'a majority, wondering if Afghanistan will ever be secure enough to return.

شهید آقا صاحب سید عظیم شاه قتالی - هرات
Shahid Sayed Azim Shah Qattali Herat.
شهید حاجی ماما عبدالعلی - هرات
Shahid Haji Ma Ma Abdul Ali Herat.
Shahid Kamalluddin Herat.

Shahid Molla Ghaffar Akhonzadah Helmand
Shahid Molla Muhammad Omar

The memories of war do not grow dimmer.

Located in the saffron-famous city, the Jihad Museum occupies a small mound. With its polished oil photographs of warlords and relics of Soviet tanks, the museum stands as a visceral reminder of what has taken place. There are empty plots next to rusted rocket launchers and destroyed tanks, as if waiting for what hasn't yet erupted.

In contrast to the Taliban's brutal rule over Hazaras in the 1990s, Naeem-ul-haq Haqqani, the thirty-seven-year-old Taliban director of Information and Culture for Herat, pledges to protect them. But in the face of the growing ISIS-K threat, Hazara mosques are too often blown to bits and their lives laid to waste.

هوايي ډګر رییس

The military side of what was once HKIA, now simply Kabul Airport, is still littered with fragments left behind in the aftermath of the frantic U.S. exodus: scattered Meals Ready-to-Eat (MREs), stationary embassy vehicles, artillery, and, most significantly, mechanics working on getting the charred aircraft back up and running.

The Taliban's top brass vow is that building an air force—the one military unit they never had as an insurgency—is a critical component of their governing agenda.

SIEMENS

د اشغال خاتمه او د خپلواکي ترلاسه کول
د ټولو هېوادوالو ګډ افتخار دی
خاتمه‌ی اشغال و کسب آزادی افتخار
مشترک ماست
Ending the occupation and gaining independence
is mutual pride of the entire nation
وزارت اطلاعات و فرهنگ

With all the confusion and crisis, many Afghan athletes—especially those who knew little of life before the U.S. occupation and came of age when women were allowed to grasp the highest echelons in sports—feel frightened and left behind.

In an outdoor café on a late fall afternoon, two sisters—Nilab Wali, twenty-six, and Anzoorat Wali, nineteen—sit somberly against the cold cushions, their eyes darting out into the fading day. The sentiment of sadness is palpable; it is one of collapsed hopes and upended lives.

The Wali sisters were rising stars in Afghanistan's Taekwondo scene, a passion cleaved from them as the Taliban stampeded to power. The martial art was much more than a hobby and potential profession; it was a lifestyle, an embodiment of just how far females had come in slicing through the society's gender predispositions, and a means of protecting their honor and independence.

"I joined Taekwondo because a woman in Afghanistan should be a fighter, be able to defend herself; that is why I started. But then I fell in love with it," Nilab recalls, her words weighted with melancholy. "Most of the time, I was going out of the house and seeing other ladies being teased. I did not want to be teased; that is why I decided I should be ready to fight them back."

Victoria's Secret
NEW YORK · LONDON · SHANGHAI

Girls and women are especially struggling with an array of mental health afflictions as the window of hope for escape narrows by the day.

"Every girl's life has changed, especially the athletes'. We cannot play again," Aryane Shamim, a twenty-three-year-old women's tennis coach in Kabul, tells me on another day.

As we speak, smoke clouds rise and mushroom in the distance, emanating from an ISIS attack nearby.

"Our hope is gone," Aryane continues, oblivious to the drama unfolding nearby. "It is very hard—we don't have a future."

Men have also been left in limbo amid dashed dreams and shattered livelihoods. That notion is perhaps no more harshly felt than in the country's once-rising martial arts arena, with some of the top stars rendered jobless and unable to leave the country.

"My dreams are destroyed. For the first twenty days after the Taliban victory, I was locked in my home and did not come out. I was sick for all those days," remembers Baz Mohammad Mubariz, one of Afghanistan's famed mixed martial arts (MMA) fighters, known in the international ring as "The Afghan Eagle." "My daughter is nine years old; now, she cannot go to school. On top of it, my revenue was from the fights, and that was how I could provide for my family. Now, the fights have stopped, and I can't train here. We don't have a future, and we're facing a very uncertain situation."

Then there is Himmat, who gazes into the distance, his bright blue eyes burnished by a terrible sadness.

"We are gay boys, but we cannot even share our problems with our families. My father or uncle will kill us. Or the Taliban will kill us," explains the Afghan, who thinks he is about twenty-one or twenty-two and refers to his love, Zubair, as his wife. "We are humans; we have rights, we have a life to live. But how?"

Homosexuality has long been deemed indecent and taboo in Afghan society. However, the past two decades offered a limited form of human rights protection—which was stripped away as the Taliban assailed to power on the sultry Sunday afternoon of August 15, 2021.

Even in times of fragility, Afghan culture upholds certain aspects of radiance that will never fade no matter what. As if barely missing a beat, wedding ceremonies resumed across the country following the dramatic change in power, although, given the economic downturn, they are a shell of their former selves.

Typically, Afghans spend their lives saving for—and paying off—marriage ceremonies. They are almost always bountiful events attended by hundreds, even thousands of guests to the tune of tens or hundreds of thousands of dollars. Kabul alone is adorned with more than two hundred elaborate wedding halls: often large, gold-trimmed venues resembling centuries-old palaces and opera houses that are lit up like flash royal residences at night.

Those halls still glow, refusing to stay pitched in barren silence, and the show must go on.

TAR

CONCLUSION

اسلامي امارت

Afghanistan is pockmarked by checkpoints and graves, of memorial flags gathering dust against a swirling fog that seems to settle on every landscape.

Too many gave too much.

I write away, nestled in my Kabul home on my last night. Twilight comes and goes, and my mind is racked with memories: from the faces I have encountered through the years and never saw again, to the streets right outside dotted with the poor trying to shield themselves from the frozen pavement with little more than a musty blanket.

But therein lies that buoyancy: what it means to be brave, fight, and endure. I am reminded that Afghanistan simmers with sweet souls who still dress in tailored suits, speak in soft, idyllic voices, read the literary greats, and want nothing to do with bombs and bullets.

Afghans will almost always give a guest in their country their last pot of green tea, and the tattered clothes off their back. At these times, I wish I were more than a journalist. It is at times like these I wish I had power.

"It is going to be okay," I respond to those who come to me and plead their cases for emergency evacuation, silently wishing that I was not telling a lie.

The truth is, none of us know.

Just because the U.S. leaves a place does not mean it is erased from the map. Afghanistan chugs on. The Afghan people still matter, their voices still exist, their stories still need to be told, and bearing witness is part of why we chose this profession in the first place.

Whether you agreed with the war or did not, we can all collectively agree that the way it has ended is as shameful as it is heartbreaking. While we cannot wipe the gore from our hands, what we can do is continue to support those left behind, empty-handed and afraid.

A new Afghanistan chapter has ignited. When you talk to many Afghans, they speak of eras depicted by lives in perpetual conflict: there was the time of the Soviets and the ensuing civil war, the time of the Taliban, and the time of the United States. Now, it's the time of the Talibs again—only it is plagued by more uncertainty, more anxiety, and a nagging worry that the lives of ordinary citizens could plunge deeper into poverty.

AFG
افغان آریانا
24

ANI STAN STAR
MYLOVE

Yet if there was bitterness about the U.S. departure, it has given way to a sad kind of acceptance.

"The U.S. is their country, so they went back. Even if they stayed one hundred years, at the end of the day, they would eventually go back," Waheedullah, who works in the bird market and thinks he is about fifty or fifty-two, says gently. "We have nowhere to go. Afghanistan is our home, and so we stay."

To stay, in many ways, is to fight, even if that means holding on to the one thing that matters most: faith.

"It is up to God," says a man who sells bananas on the street, but barely makes a sale a day. "He will protect us."

The Taliban banned kite-flying during its first rule. Yet as of now, children still fly kites from high on the hilltop, shrieking in delight as the nylon floats off toward the sunshine as if it was the most magnificent thrill in the world.

All I can think to say, hoping those small boys could hear me, is "fly on."

We will not say goodbye to Afghanistan.

I ♥ KABUL

www.ingramcontent.com/pod-product-compliance
Lightning Source LLC
Chambersburg PA
CBHW041008251026
53749CB00069B/74